Mehdi Ostadhassan · Bodhisatwa Hazra

Advanced Methods in Petroleum Geochemistry

Mehdi Ostadhassan
Institute of Geosciences, Marine and Land
Geomechanics and Geotectonics
Christian-Albrechts-Universität
Kiel, Germany

Bodhisatwa Hazra
Department of Coal-Rock Chemistry
and Structure
Central Institute of Mining and Fuel
Research
Dhanbad, Jharkhand, India

ISSN 2509-3126   ISSN 2509-3134 (electronic)
SpringerBriefs in Petroleum Geoscience & Engineering
ISBN 978-3-031-44404-3   ISBN 978-3-031-44405-0 (eBook)
https://doi.org/10.1007/978-3-031-44405-0

This Springer imprint is published by the registered company Springer Nature Switzerland AG
The registered company address is: Gewerbestrasse 11, 6330 Cham, Switzerland

Paper in this product is recyclable.

SpringerBriefs in Petroleum Geoscience & Engineering

The SpringerBriefs series in Petroleum Geoscience & Engineering promotes and expedites the dissemination of substantive new research results, state-of-the-art subject reviews and tutorial overviews in the field of petroleum exploration, petroleum engineering and production technology. The subject focus is on upstream exploration and production, subsurface geoscience and engineering. These concise summaries (50-125 pages) will include cutting-edge research, analytical methods, advanced modelling techniques and practical applications. Coverage will extend to all theoretical and applied aspects of the field, including traditional drilling, shale-gas fracking, deepwater sedimentology, seismic exploration, pore-flow modelling and petroleum economics. Topics include but are not limited to:

- Petroleum Geology & Geophysics
- Exploration: Conventional and Unconventional
- Seismic Interpretation
- Formation Evaluation (well logging)
- Drilling and Completion
- Hydraulic Fracturing
- Geomechanics
- Reservoir Simulation and Modelling
- Flow in Porous Media: from nano- to field-scale
- Reservoir Engineering
- Production Engineering
- Well Engineering; Design, Decommissioning and Abandonment
- Petroleum Systems; Instrumentation and Control
- Flow Assurance, Mineral Scale & Hydrates
- Reservoir and Well Intervention
- Reservoir Stimulation
- Oilfield Chemistry
- Risk and Uncertainty
- Petroleum Economics and Energy Policy

Contributions to the series can be made by submitting a proposal to the responsible Springer contact, Anthony Doyle at anthony.doyle@springer.com.

Preface

Petroleum geochemistry is a vital domain in the exploration of oil and gas resources, as it involves in-depth analyses of source rocks and the fluids they produce. By applying general chemistry principles, petroleum geochemistry delves into the intricate processes of origin, generation, migration, accumulation, and alteration of petroleum found in organic-rich fine-grain sedimentary rocks. Traditionally, geochemistry methods have relied on bulk samples to provide insights at a macro or basin scale. However, the surge in unconventional shale plays has exposed the limitations of these conventional methods, especially in meeting our needs for nano to micron scale information. The heterogeneity of these resources leads to significant variations in chemistry and formation attributes, occurring at an exceptionally fine scale from source to reservoir rocks, given their close proximity. While conventional methods are adequate for understanding hydrocarbon generation, migration, and accumulation in a broader context, they fall short when it comes to examining finer-scale characteristics.

Addressing these challenges is paramount for advancing our understanding of petroleum systems. To overcome these limitations, researchers must embrace advanced analytical instrumentation, including cutting-edge techniques like Raman spectroscopy, atomic force microscopy-infrared (AFM-IR), nuclear magnetic resonance (NMR), and mass spectroscopy. These innovative approaches open new possibilities for unveiling intricate details and resolving ambiguities in unconventional resources. By incorporating these advanced methods, petroleum geochemistry can provide robust answers to the remaining questions in the field and enable comprehensive evaluations of these valuable energy resources.

In this context, the presented monograph proposes novel methods that have recently emerged in the realm of petroleum geochemistry, with a primary focus on the evaluation of source rocks. As a crucial aspect of organic geochemistry, source rock analysis involves examining organic-rich fine-grain sediments for their thermal maturity, production potential, chemical structure evolution during burial history, and migration patterns. Traditionally, this type of analysis has mainly relied on bulk rock samples, neglecting finer-scale studies at the nano or micro levels. However, with the rise of unconventional resources like shale oil and gas, where organic-rich fine-grain sediments act as both reservoir and source rocks, and hydrocarbons migrate over short distances, conventional methods like programmed pyrolysis prove insufficient to meet the evolving needs of petroleum geochemists and geologists.

In response to these evolving demands, the monograph unveils cutting-edge techniques that go beyond traditional petroleum geochemistry procedures, offering a more comprehensive understanding of petroleum systems. These innovative methods, such as Raman spectroscopy, NMR, AFM-IR, and XPS spectroscopy, are based on advanced and recently developed instrumentation, enabling a multi-scale approach to studying petroleum systems. The monograph seamlessly integrates these unconventional analytical techniques with traditional methods, demonstrating how they can be correlated to reveal a wealth of information at different scales, from the nano to the macro level. Throughout the book, the authors generously provide data, images, and detailed explanations of the methods, data collection, interpretation of results, and intercorrelated characteristics of the study specimens.

By embracing this amalgamation of advanced and traditional techniques, petroleum geochemists can unlock new insights into the mechanisms of petroleum generation, expulsion, migration, and accumulation in reservoirs from the source rock. The integration of these innovative analytical tools paves the way for a more profound and comprehensive understanding of petroleum systems and their inherent complexities, ultimately shaping the future of energy exploration and exploitation.

Kiel, Germany   Mehdi Ostadhassan, Ph.D.

Contents

1 Molecular Heterogeneity of Organic Matter in Geomaterials by AFM Based IR Spectroscopy 1

   1.1 Introduction 1

   1.2 Principle of Atomic Force Microscopy (AFM) 2

   1.3 AFM Integration with-Infrared (AFM-IR) Spectroscopy 4

      1.3.1 Case Study 7

   1.4 Detection of OM Heterogeneity 8

   1.5 Conclusion 25

   References 25

2 A Chemo-mechanical Snapshot of In-Situ Conversion of Kerogen to Petroleum 27

   2.1 Introduction 27

   2.2 Evolution of Chemical Composition 32

   2.3 Nanomechanical Properties 38

   2.4 Conclusion 40

   References 40

3 Bacterial Versus Thermal Degradation of Algal Matter: Analysis from a Physicochemical Perspective 43

   3.1 Introduction 43

   3.2 Organic Petrology 47

   3.3 Chemical Mapping 48

   3.4 Nanomechanical Mapping 52

   3.5 Conclusion 56

   References 57

4 Understanding Organic Matter Heterogeneity and Maturation Rate by Raman Spectroscopy 61

   4.1 Introduction 62

   4.2 Raman Spectroscopy 64

   4.3 Evolution of OM Thermal Maturity Pathways 68

4.4 OM Heterogeneity 73
4.5 Conclusion 80
References 81

**5 Backtracking to Parent Maceral from Produced Bitumen
with Raman Spectroscopy** 87
5.1 Introduction 87
5.2 Raman Spectroscopy and Maturity 91
5.3 Raman Spectroscopy and Kerogen Typing 94
5.4 Conclusion 99
References 100

**6 Structural Evolution of Organic Matter in Deep Shales
by Spectroscopy (^{1}H & ^{13}C-NMR, XPS, and FTIR) Analysis** 105
6.1 Introduction 106
6.2 Materials and Methods 107
     6.2.1 Solid-state ^{1}H & ^{13}C-NMR (Nuclear Magnetic
            Resonance) 107
     6.2.2 XPS (X-Ray Photoelectron Spectroscopy) 108
     6.2.3 FTIR (Fourier Transform Infrared Spectroscopy) 110
6.3 Results 110
     6.3.1 Solid-State ^{1}H & ^{13}C-NMR 110
     6.3.2 XPS 112
     6.3.3 FTIR 113
6.4 Discussion 115
     6.4.1 Carbon Structural Changes 115
     6.4.2 Heteroatoms 117
6.5 Conclusions 119
References 120

Chapter 1
Molecular Heterogeneity of Organic Matter in Geomaterials by AFM Based IR Spectroscopy

Abstract Several samples were selected at the early and peak thermal maturity stages of organic matter, based on bulk geochemical screening, organic petrology and fluorescence emission of the liptinite group maceral and solid bitumen reflectance. Identified particular organic material in the samples were examined by AFM-IR spectroscopy to evaluate organic matter heterogeneity at the nanoscale, based on chemical variations. A significant chemical heterogeneity was observed within unaltered telalginite and bacterial degraded *Tasmanites*, and also between two separate solid bitumens that are next to one another and at the same stage of thermal progression. Furthermore, considering their separate pathways of generation, these solid bitumen particles were compared in terms of their chemical content. While thermal maturity progression was found to reduce molecular chemical heterogeneity in the organic matter particles during the maturation pathway, on the contrary, during the bacterial degradation, the *Tasmanites* has lost its fluorescence emission and the relative chemical heterogeneity was increased compared to the unaltered telalginite, a phenomenon that was observed for the first time.

Keywords Bakken shale · Thermal maturity · Organic matter heterogeneity · AFM based nano-IR spectroscopy · Organic petrology

1.1 Introduction

Considering shale as a source rock, the organic matter (OM) plays a critical role in increasing the degree of heterogeneity of the rock. For instance, where the organic matter is mostly solid bitumen (SB), overall reservoir quality can get affected [25, 32] leading to a considerable decrease in the permeability and pore throat sizes [21]. Also, the submicron length scale heterogeneity of OM in shale has been demonstrated by SEM observations, where it has occasionally been found that adjacent OM grains have significantly different porosities. This heterogenous nature of the OM may result from inherent variability in the OM caused due to factors like the depositional environment as well as potential regional differences in the degree of thermal

M. Ostadhassan and B. Hazra, *Advanced Methods in Petroleum Geochemistry*,
SpringerBriefs in Petroleum Geoscience & Engineering,
https://doi.org/10.1007/978-3-031-44405-0_1

alteration brought about by catalysis from mineral grains present in close proximity [33]. Assessing shale in multiscale and OM with respect to physio-chemical heterogeneities helps gaining better insight into reservoir performance and mechanisms that would lead to generation of petroleum from the OM [8]. While SEM offers thorough, high-resolution details on OM pores, it offers no details on their type or chemical makeup. Reflected light optical microscopy is employed to distinguish between various OM types, but it does not offer molecular data. Rock–Eval and infrared spectroscopy are two methods that have been used to assess changes in the average chemical composition of OM at various thermal maturities, but they are unable to detect geochemical heterogeneity in shales at low spatial resolution. Although Fourier transform infrared microscopy (micro-FTIR) is used to study this heterogeneity, the diffraction limit prevents it from reliably resolving chemical characteristics at the sub-micron length scale pertinent to shale [7, 33]. The extent of organisation of OM at the micron length scale can also be determined structurally using Raman imaging, but this method is limited by the strong fluorescence background from immature shale samples. In addition to topographic imaging by AFM, AFM-IR, a rapidly developing technique in the materials and life sciences, offers chemical and modulus mapping at the nanoscale that is not affected by the diffraction limit. Yang et al. [33] for the first-time applied AFM-IR to determine the heterogeneity in shales in terms of their chemical and mechanical properties at nanometre scale spatial resolution. They quantified nano chemo-mechanical characteristics of different types of OM (solid bitumen, inertinite, and *Tasmanites* telalginite) on artificially matured (hydrous pyrolysis) samples from the New Albany Shale. They noted that while each maceral type's average composition varies significantly, there is little variation in the composition between different locations that belong to the same maceral type. AFM based nanoIR technique, as was employed by Yang et al. [33] can map chemical variations at the nanoscale without getting affected by the diffraction limit.

1.2 Principle of Atomic Force Microscopy (AFM)

Different methods for characterising rocks are chosen at various resolutions [18]: (i) seismic imaging at resolutions of tens of metres, (ii) well logging and drilling measurements at resolutions of less than 1 m, and (iii) mechanical and petrophysical properties of rocks (permeability, porosity, and capillarity), which can be measured at a resolution of centimetres in core tests. Properties of rocks can be highly heterogenous even at the nanoscale resolution [15]. Thus, nano scale characterization can be a useful tool to improve rock characterization. Atomic force microscopy (AFM) and nanoindentation, generally called as force spectroscopy, are commonly used for micro/nano scale measurement on a wide range of materials to assess chemical, electrical and mechanical properties of various types of biological system, nanoscale materials as well as geomaterials. Although scanning electron microscopy (SEM) can be used to characterise rocks at the nanoscale, it does so by emitting and collecting photons and electrons in a manner similar to our sense of light, whereas AFM uses a stylus tip to scan a sample in a manner similar to our sense of touch [18]. The AFM measures interactive forces between surfaces directly and can be used for:

(a) surface topography measurement
(b) simultaneous determination of the surface materials
(c) determining the local bulk modulus of elasticity of the rock material for use in geomechanical models for hydraulic-fracture design.

Binnig et al. [5] introduced the first AFM model as a scanning-tunneling microscope (STM). A tunnelling current that is inversely proportional to the gap width is produced in STM by bringing a voltage-biased metal tip close to the surface. By moving the tip vertically while scanning the surface, the STM feedback system maintains a constant tunnelling current. The piezoelectric scanner determines the exact x, y, and z positions of the tip. The vertical position provides surface topography because a constant current guarantees a constant gap width. However, specific STM principles only allow imaging of conducting or semiconducting surfaces. Subsequently, Binnig et al. [6] suggested the tip to be mounted on the cantilever spring and observing the deflection of the cantilever relative to surface/tip forces, thereby creating the AFM. Since the force between the surface and the tip is dependent on the width of the gap, feedback system of the AFM can keep an even tip/surface gap by compensating for cantilever deflection with a vertical tip displacement. Imaging the topography of any material, whether conducting or nonconducting, is possible with this method.

Figure 1.1 shows schematic of a typical AFM setup consisting of a cantilever with a tip attached to its end (tip nose radius is 30 nm), a laser source, a chip holder, quadrant photodiode, controlling system, and mirror. The cantilever moves at the prescribed speed vertically with the help of piezoelectric movement. The deflection of cantilever in the z-direction as the tip nears the surface and retracts from it is recorded by the AFM so that it can directly measure the interactive forces between the surfaces (i.e., tip nose and substrate). The cantilever can be modelled as a spring because one end is held, and the other is free. The force between the tip surface and the cantilever can then be calculated using Hooke's law:

$$F = K_c Z_c, \tag{1.1}$$

where "Z_c" denotes cantilever deflection at its free end, "k_c" denotes cantilever stiffness or spring constant, and "F" denotes the force acting between the tip and the surface. In Fig. 1.2a, a tip-cantilever is shown approaching and touching a surface before retracting from the surface in a schematic representation of the approach and the retract movements. In the stages of approach and retract, force responses related to tip/surface interactions are shown in Fig. 1.2b as a function of distance. During the initial phase of the approach, the cantilever simply shows no deflection and detects no interactive force. When the tip gets close to the surface, van der Waals (vdW) forces of attraction start to work and b end the cantilever in the direction of the surface. The tip rises to the surface as soon as the cantilever-spring constant is exceeded by the

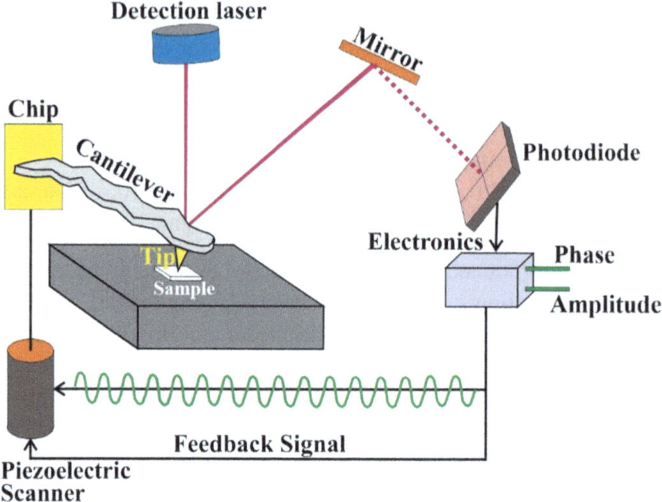

Fig. 1.1 Schematic of AFM scanner-head system. Modified after Javadpour et al. [18]

gradient of vdW forces. After that, the tip comes into contact with the surface, and if it is advanced any further, leaves an indentation on the sample. The tip typically maintains contact with the sample during retraction due to adhesion after the zero force has passed. A hysteresis loop is typically formed as a result of adhesive forces, plastic deformation, and/or an increase in the area of contact. The tip then detaches from the surface due to the chip's vertical movement, and force interaction goes back to the original no-interaction phase.

1.3 AFM Integration with-Infrared (AFM-IR) Spectroscopy

Both AFM and infrared spectroscopy are potent and frequently used methods on their own. AFM is frequently employed for topographic imaging of a wide range of samples in materials and life science research in addition to numerous industrial uses. On the other hand, infrared spectroscopy is popular technique for chemical analysis. Chemical characterization is carried out using infrared spectroscopy, which measures the amount of infrared light that a specimen absorbs as a function of the frequency (or equivalently wavelength) of the IR light. Chemical species can be characterised and/or identified using the form of absorption peaks in the IR spectra, which act as a fingerprint.

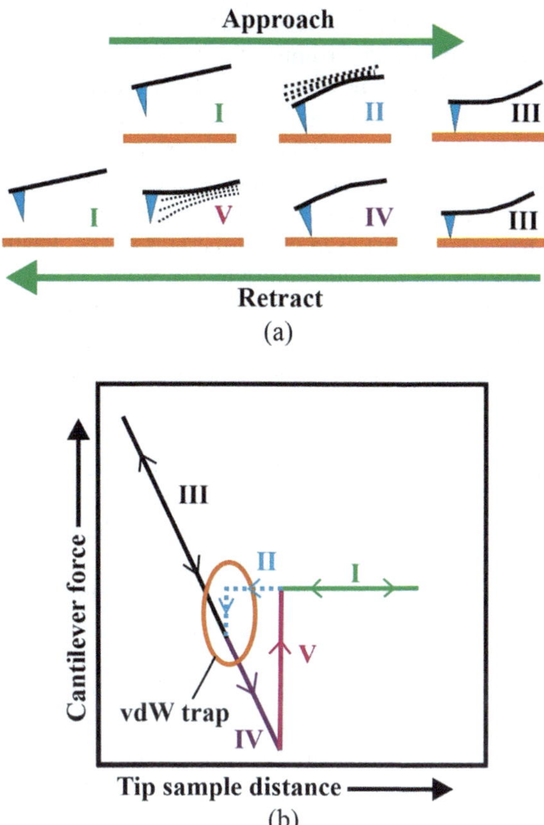

Fig. 1.2 a Different phases occur as a tip approaches a surface (I-III) and retracts from the substrate (III-I). Step A of the approach happens when the tip is far from the surface and there is no contact between the surface and the tip. The tip interacts with the surface in step B by getting close to it. The tip is given a small push once it makes contact with the surface (Step III). The tip retracts from the surface at Step III of the retraction phase, but the adhesive forces cause it to stay attached (Step IV). Step V is where the tip abruptly separates from the surface and returns to its initial position (i.e., Step I); **b** The symbolic force curve that corresponds to the process in **a**

AFM-IR is a combination of two techniques to reap the benefits of both AFM in terms of its spatial resolution capacity and infrared spectroscopy in terms of its chemical analysis capability. AFM can consistently accomplish spatial resolution of nanometers within the resolution of the probe tip of AFM, but for chemical analysis it is not useful [29]. Similarly, despite widespread use of Fourier transformed infrared spectroscopy for chemical characterization of coal, shale, minerals, and microfossils [29], infrared microscopes have limitations due to optical diffraction and frequently used IR sources of low brilliance. The fundamental limit for spatial resolution that optical diffraction typically imposes is $\lambda/2$, where λ is the illumination wavelength. On the basis of the specific configuration and technique used, the majority of commercial FT-IR microscopes using thermal infrared sources have a functional spatial resolution limit of λ to 3λ, attaining spatial resolution ranging between 2.5 and 75 µm. IR beamlines of high-brilliance synchrotron, where performance limited by diffraction has been attained, have been coupled to FT-IR microscopes. However, most of the industrial and research characterization issues on the length scale of many micrometres and above have been the focus of conventional infrared micro spectroscopy applications. Recent developments in array and optics detectors have shown that even with a thermal source, FT-IR microscopy is capable of achieving spatial resolution close to 1 m scale. The AFM-IR technique involves examining the local transient deformation brought on by the photothermal expansion effect. It has had great success in the fields of materials and life sciences and can get around the two main aforementioned drawbacks of AFM and infrared microspectroscopy.

An infrared laser beam is used to illuminate the AFM probe after it has been positioned above a region of interest in an AFM-IR setup (Fig. 1.3). The cantilever of the AFM oscillates because the surface of the sample experiences thermal expansion when the wavelength of the laser falls within a particular infrared absorption band. The resulting force impulse's brief duration induces a number of oscillating modes in the cantilever, and in most instances, the flux amplitude is proportionate to the regional infrared absorption [9]. A sample's mechanical stiffness can be determined from the contact resonance frequency. As a result, the coupled tip-sample contact system can quantitatively obtain both the stiffness and local infrared absorption images at the same time. Since frequencies of contact resonance vary with the elasticity of sample, information about chemical composition as well as structural characteristics is also provided [9].

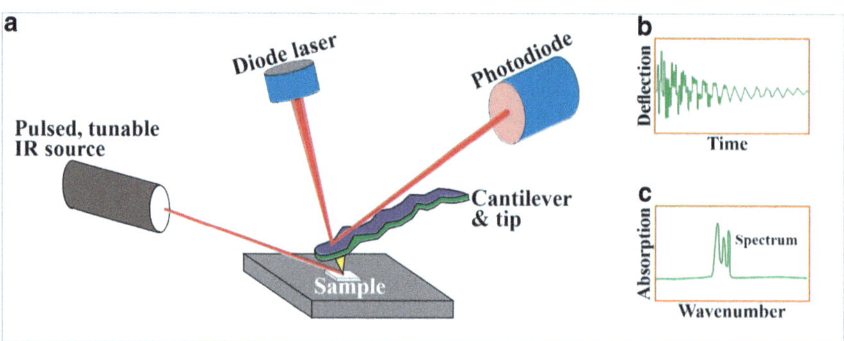

Fig. 1.3 Schematic of AFM-IR: **a** schematic diagram of AFM-IR; **b** IR absorption proportional to cantilever oscillation **c** cantilever oscillation and wavelength, compared to obtain absorption spectrum

1.3.1 Case Study

Abarghani et al. [3] used AFM-IR spectroscopy in conjunction with organic petrography to evaluate the influence of thermal maturity on heterogeneity of OM in four individual macerals at two stages (early mature and peak mature) of natural thermal maturity for the first time. The selected Peak mature samples were solid bitumen and bituminized *Tasmanites* and early mature samples were solid bitumen, telalginite, and degraded *Tasmanites*. They selected these samples to demonstrate that there is a significant level of heterogeneity not only among these macerals but also within each individual maceral as well. Their study also noted chemical alterations between bacterial degradation and natural thermal maturity pathways.

They selected three shale samples from the Lower Bakken Formation in the state of North Dakota, a major unconventional shale play, through bulk geochemical screening (programmed pyrolysis). The Bakken Formation (source rock portion) is an exceptional rock unit for geochemical studies since all stages of thermal maturity (from immature to relatively postmature) [30] can be found in one location. Previous basin-wide detailed investigations [1, 2, 14] of shale members demonstrated that the OM consists mostly of kerogen Types II and mixed II/III. The constituent macerals and OM is comprised of solid bitumen, amorphous matrix bituminite (hebamorphinite and fluoramorphinite), unicellular marine telalginite (*Tasmanites*, *Leiosphaeridia*, and Prasinophytes), unstructured algal fragments, sporinite, acritarchs, amorphinite, liptodetrinite, granular micrinite, macrinite, inertinite, and minor zooclast-like fragments.

Table 1.1 represents the results from open-system temperature-programmed pyrolysis-oxidation experiments conducted by Abarghani et al. [3]. The total organic carbon (TOC%) of sample 1 and 2 was observed to be 14.96% and 17.64%, respectively. Sample 2 was observed to be early mature [23] having T_{max} (temperature maxima of the S_2 curve generated from decomposition of kerogen during pyrolysis generating heavier hydrocarbons) of 437 °C, hydrogen index (HI = $(S_2/TOC) * 100$; as a proxy for the atomic H/C ratio of kerogen) of 534 mg HC/g TOC, and production index (PI-Espitalie et al. [11]) of 0.10 (Table 1.1). This confirmed the early stage of the oil window, which was also supported by the fluorescence colors of telalginite (*Tasmanites*) that varied from pale greenish-yellow to golden-yellow (Fig. 1.4d). Furthermore, the very low OI of 2 mg CO_2/g TOC (Table 1.1) suggested a relatively high H/C ratio and low O/C ratio for this sample. A solid bitumen particle within Sample 2 ($BR_O \approx 0.40\%$; N = 52 and SD = 0.081) showing no visible fluorescence, was selected by them for further examination (Fig. 1.4a, b). A combination of T_{max} of 436 °C, HI of 557 mg HC/g TOC, and PI of 0.07 (Table 1.1) confirmed the early stage of the oil window for sample 1. This sample also exhibited low OI of 2 mg CO_2/g TOC, indicating relatively high H/C ratio and low O/C ratio. Pale greenish-yellow to golden-yellow fluorescence colors of the telalginite macerals across sample 1 and 2 also confirmed the early mature stage for the OM of these two specimens. In

Table 1.1 Organic geochemical parameters of three bulk samples from the lower shale member of the Bakken formation

Sample ID	Weight	S_1	S_2	T_{max}	S_3	TOC	HI	OI	PI
	Mg	mg HC/g	mg HC/g	°C	mg CO_2/g	wt%	mg HC/g TOC	mg CO_2/g TOC	–
1	61.0	6.19	83.26	436	0.25	14.96	557	2	0.07
2	60.7	10.76	94.24	437	0.35	17.64	534	2	0.10
3	61.0	6.06	47.74	445	0.36	15.27	313	2	0.11

the case of sample 3, the TOC was observed to be 15.27 wt%. T_{max} of 445 °C, HI of 313 mg HC/g TOC, and PI of 0.11 (Table 1.1) as well as the dull golden-yellow and, in part, light-orange fluorescence color of the *Tasmanites* telalginite under UV light confirmed that the OM is mature and in the middle stage/peak of the oil window. In their study, Abarghani et al. [3] observed two different generations of solid bitumen present side-by-side in the samples (Fig. 1.5a). The larger particle on the left was solid bitumen with mean R_O, ran of 0.38% (N = 20) and did not show any fluorescence. The smaller particle on the right is in-situ solid bitumen, which is the product of the in-situ bituminization of the *Tasmanites* telalginite. In-situ conversion of telalginite to bitumen has already been documented [21]. During this process, telalginite (the parent maceral) is gradually converted into bitumen in-situ without any hydrocarbon migration. The bituminized telalginite initially shows fluorescence, while the intensity of UV emission is lost when a complete conversion of telalginite to solid bitumen is achieved. In this example, bitumen still exhibited the dull golden-yellow fluorescence inherited from the parent telalginite, which suggested that the bituminization of the *Tasmanites* did not finish. The mean R_O ran of the bituminized *Tasmanites* was 0.29% (N = 20). The R_O% map of the particle on the right showed a greater variability compared to the particle on the left (Fig. 1.5c) (Standard Deviation of 0.064 in the bituminized *Tasmanites* versus 0.025 in the solid bitumen) which suggested separate physio-chemical pathways in each particle.

1.4 Detection of OM Heterogeneity

In their study, Abarghani et al. [3] AFM-based Nano-IR based spectroscopy on targeted ROIs marked in the individual macerals (Figs. 1.4a and 1.5a, c) to delineate chemical variations and reveal organic matter heterogeneity in nanoscale. Although organic petrology was unable to reveal the heterogeneity of the organic matter in molecular scale, it still provided supporting information about the OM heterogeneity (Figs. 1.4b and 1.5c) to foster physio-chemical properties across the surface of the OM particles.

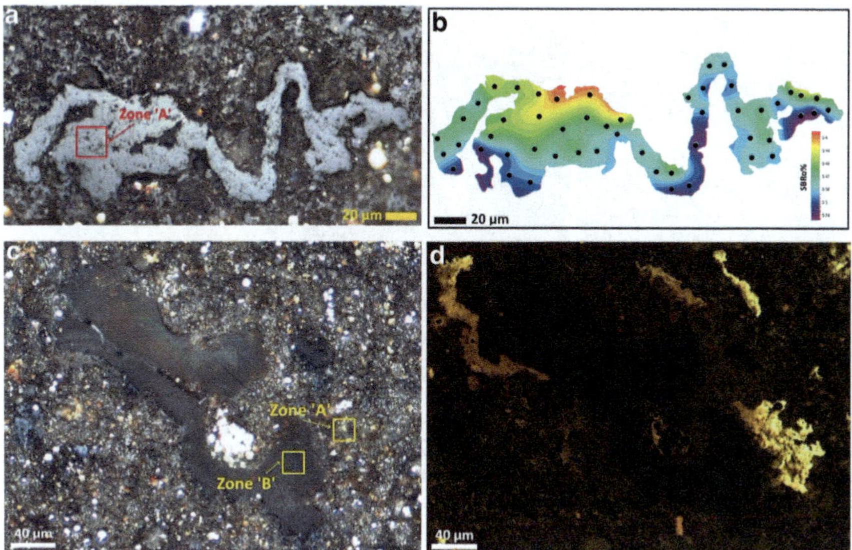

Fig. 1.4 **a** Photomicrograph of solid bitumen in sample 2 under white incident light. This particle has no visible fluorescence. Zons 'A' was selected for AFM-based IR spectra measurements. **b** $SBR_O\%$ reflectance map for the same particle based on 52 data points shown as black dots. **c** Photomicrograph of the degraded Tasmanites (telalginite) in sample 1 based on morphology and internal reflections (the 'peacock' colors seen in the upper part) and a slightly grainy texture that appears in patches within the degraded telalginite under white incident light. Zones 'A' and 'B' were selected for AFM-based IR spectra measurements. **d** Same view as 2C under UV light. Telalginite exhibits pale greenish-yellow to golden-yellow fluorescence colors, suggesting early mature stage. The degraded Tasmanites does not exhibit any fluorescence probably due to the effect of degradation. The enclosing amorphous bituminite matrix fluoresces with olive-green color

A Peak mature sample

Two zones (ROI) were selected for chemical mapping in the peak mature sample (Fig. 1.5a), one trajectory through the boundary between two macerals (Fig. 1.5a: Zone A) and another one (Fig. 1.5a: Zone B) inside the non-fluorescing solid bitumen (Fig. 1.5b). They acquired several IR spectra inside a $20 \times 20\ \mu$m area (Fig. 1.6) in order to investigate chemical variations inside and between these two different types and generations of solid bitumen for comparison. AFM height maps showed relatively smooth surfaces (roughness less than 340 nm) for both particles (Fig. 1.6a, b) while IR absorption mapping of aliphatic C–H stretching (Fig. 1.6c, d: $2920\ \mathrm{cm}^{-1}$) uncovered chemical heterogeneity across the surveyed zones, especially for the bituminized *Tasmanites* particle (Fig. 1.6d: Zone A). They collected IR spectra at eleven data points on the solid bitumen surface and four points on the bituminized *Tasmanites* (Fig. 1.6a, b). The following spectral ratios were computed by Abarghani et al. [3] for the chemical interpretation of each data point (Table 1.2):

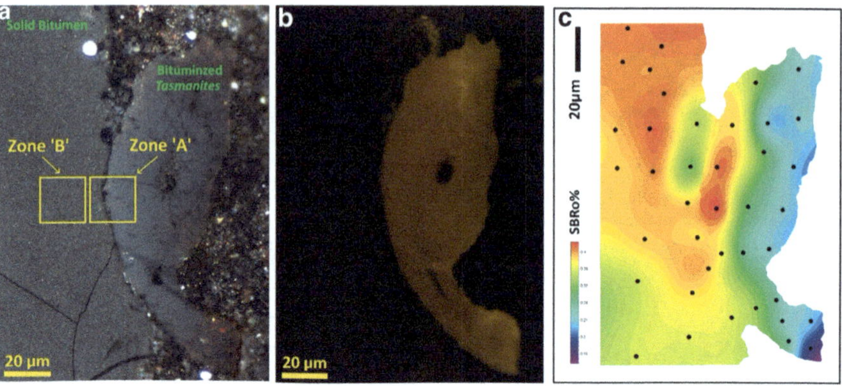

Fig. 1.5 a Solid bitumen particles in sample 3. Two different generations of bitumen, the right one is comprised of in-situ bituminized Tasmanites. Zones 'A' and 'B' were selected for AFM-based IR spectra measurements. Photomicrograph was taken using a 50× oil immersion objective under white incident light. **b** Same view as 1A under UV light. The bituminized Tasmanites is showing dull golden-yellow and partly light-orange fluorescence color. **c** $SBR_O\%$ reflectance map for the studied particles based on 40 data points shown as black dots. The bituminized Tasmanites shows a greater variability as a function of $SBR_O\%$

1. CH_3/CH_2 represents the C–H region, aliphatic stretching [22]. Different methods are proposed for calculating this index based on symmetrical CH_2 and CH_3 stretching (e.g., 2850 and 2866 cm^{-1}, respectively) or asymmetrical CH_2 and CH_3 stretching (e.g., 2920 or 2925, and 2955 cm^{-1}, respectively) [19, 20]. Here, we used the asymmetrical stretching of 2958 and 2919 cm^{-1} for computing this ratio. A decrease in CH_3/CH_2 ratio represents longer and less branched aliphatic chains [19].

2. Ali/Ox indices as the ratio of the aliphatic stretch (2800–3000 cm^{-1}) to the oxygenated functions (1500–1800 cm^{-1}).

3. 'A' and 'C' factors [13] that are defined as the ratios of (2930 cm^{-1} + 2860 cm^{-1})/ (2930 cm^{-1} + 2860 cm^{-1} + 1630 cm^{-1}), and (1710 cm^{-1})/(1710 cm^{-1} + 1630 cm^{-1}), respectively. These factors consider absorption peaks of aliphatic CH_2 and CH_3 (at 2860 and 2930 cm^{-1}), carboxyl and carbonyl groups (at 1710 cm^{-1}), and aromatic C=C bonds (at 1630 cm^{-1}). Ganz and Kalkreuth [13] argued that these factors can replace the traditional Van Krevelen diagram ('A' as H/C, and 'C' as O/C ratios) when interpreting maturity trends and kerogen types.

4. AR H3000–3100 cm^{-1}/AL 2800–3000 cm^{-1}, representing the ratio of the aromatic C–H stretching to the aliphatic C–H stretching region.

5. AR H3000–3100 cm^{-1}/AL 1450 cm^{-1}, CH_2, and CH_3 absorption bands that may also incorporate some aromatic rings (1450 cm^{-1}) [7].

6. AR H3000–3100 cm^{-1}/AL 1370 cm^{-1}, CH_3 and C–H bending in the methylene group (1370 cm^{-1}).

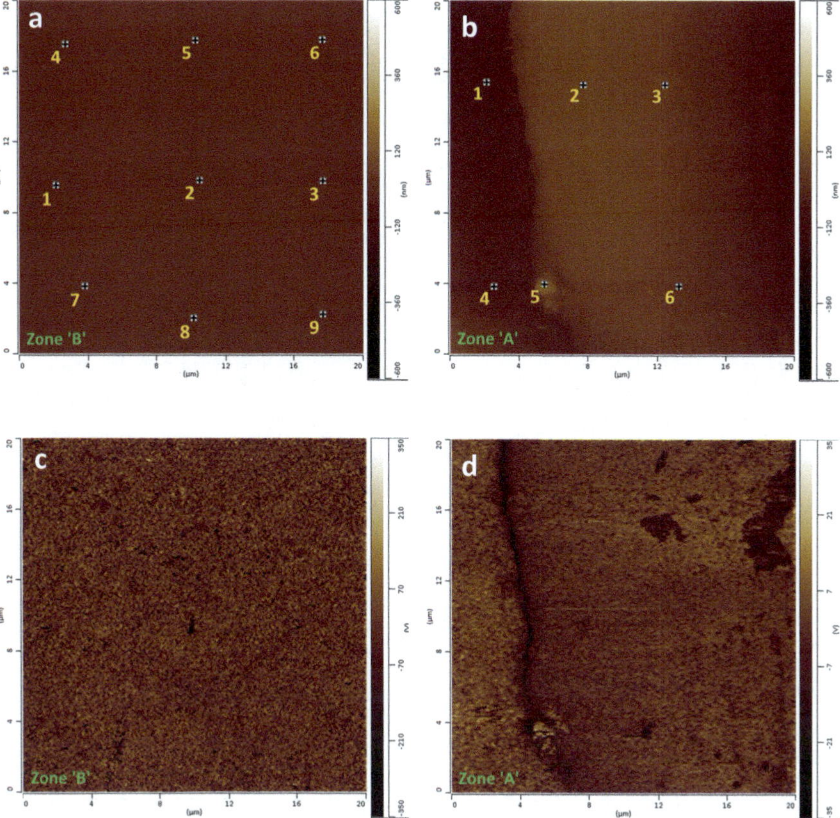

Fig. 1.6 **a** and **b** AFM height map (topography) image of zones 'A' and 'B' showing locations of individual spectral data points labeled 1–9 inside the solid bitumen, and 1–6 at the edge part of the solid bitumen and inside the bituminized Tasmanites. **c** and **d** Same view as in **a** and **b** showing IR chemical mapping of aliphatic C–H stretching (2920 cm^{-1}) across zones 'A' and 'B' in the solid bitumen and in the bituminized Tasmanites, respectively

7. AR C1600 cm^{-1}/AL 2800–3000 cm^{-1}, olefinic/aromatic ring stretch (1600 cm^{-1}) [7, 31].
8. AR C1600 cm^{-1}/AL 1450 cm^{-1}, and finally,
9. AR C1600 cm^{-1}/AL 1450 cm^{-1}.

Abarghani et al. [3] noted that solid bitumen had a relatively lower degree of chemical heterogeneity compared to the bituminized *Tasmanites* (Fig. 1.6b: Zone B). However, they still observed some variation in functional groups in the bandwidths of 1000–1150, 2850–2900, and 3000–3650 cm^{-1}. Therefore, they compared these two different types of bitumen by taking of average of the spectral intensity of each bitumen (Fig. 1.7). It was observed that the most apparent disparity between the two types of bitumen became clear at the 2800–3000 cm^{-1}, and 3000–3100 cm^{-1} bandwidths representing aliphatic C–H and aromatic C–H

Table 1.2 The CH_3/CH_2, Ali/Ox, 'A' and 'C' factors, and ratios of aromatic versus aliphatic Nano-IR absorption values for the peak mature sample

Sample	Spectra	CH_3 CH_2	Ali Ox	A_{Factor}	C_{Factor}	$\dfrac{AR_{H3000-3100}}{AL_{2800-3000}}$	$\dfrac{AR_{H3000-3100}}{AL_{1450}}$	$\dfrac{AR_{H3000-3100}}{AL_{1370}}$	$\dfrac{AR_{C1600}}{AL_{2800-3000}}$	$\dfrac{AR_{C1600}}{AL_{1450}}$	$\dfrac{AR_{C1600}}{AL_{1370}}$
Bituminized *Tasmanites*	2	0.87	0.86	0.65	0.19	0.49	14.06	7.21	0.61	17.51	8.99
	3	0.50	0.91	0.91	0.59	0.36	7.99	4.41	0.34	7.46	4.12
	5	0.73	0.94	0.74	0.82	0.46	7.46	3.68	0.31	5.11	2.52
	6	0.70	1.02	0.23	0.28	0.38	7.85	8.78	0.12	2.50	2.80
	Average	0.70	0.93	0.63	0.47	0.42	9.34	6.02	0.35	8.15	4.61
	St. Dev	0.16	0.07	0.29	0.29	0.06	3.15	2.39	0.20	6.57	3.00
SB (margin)	1	0.86	0.71	0.55	0.32	0.57	1.35	4.30	0.51	1.19	3.79
	4	0.04	0.97	0.86	0.29	0.48	1.75	8.69	0.22	0.81	4.00
Solid bitumen	1	0.65	0.91	0.70	0.22	0.38	2.97	2.20	0.19	1.45	1.07
	2	0.41	0.81	0.58	0.63	0.39	6.66	5.21	0.79	13.70	10.73
	3	0.24	0.82	0.43	0.55	0.36	6.76	2.47	0.75	14.19	5.19
	4	0.87	0.79	0.81	0.37	0.34	2.52	3.16	0.50	3.78	4.75
	5	0.12	0.75	0.38	0.12	0.32	5.83	1.79	0.48	8.66	2.66
	6	0.19	0.92	0.59	0.15	0.35	7.66	1.60	0.11	2.36	0.49
	7	0.17	0.96	0.54	0.45	0.39	7.44	2.18	0.33	6.19	1.81
	8	0.49	0.85	0.46	0.13	0.40	5.63	1.28	0.67	9.45	2.16
	9	0.99	0.85	0.93	0.81	0.33	2.21	2.90	0.65	4.35	5.70
	Average	0.46	0.85	0.62	0.37	0.39	4.61	3.25	0.47	6.01	3.85
	St. Dev	0.34	0.08	0.18	0.22	0.07	2.45	2.15	0.24	4.87	2.84

stretching, respectively [20], where the solid bitumen exhibited a lower AR H3000–3100 cm^{-1}/ AL 2800–3000 cm^{-1} ratio (0.36 vs. 0.42, respectively from the averaged spectra, Table 1.3). On consideration of other aliphatic functional groups, they noted that the solid bitumen showed higher values of integrated peak areas from AL 1450, AL 1370 cm^{-1}, but lower values of AL 1000–1100 (aliphatic ethers, alcohols), compared to the bituminized *Tasmanites*. Aromatic functional groups including AR 1490 cm^{-1} and AR 1600 cm^{-1} exhibited more concentration in the bituminized *Tasmanites*, where higher values of AR 1600, AR 1490, and the aromaticity index (AR H3000–3100/AL 2800–3000) were observed compared to the solid bitumen (Table 1.3). Lis et al. [20] advocated for AL H2800–3000 cm^{-1} as the most appropriate bandwidth for quantifying aliphatic absorption. They stated that the accuracy of absorption at AL 1450 and AL 1370 cm^{-1} bandwidths is limited due to analytical errors. Furthermore, as observed in Abarghani et al. [3], the two types of bitumen illustrated considerable variations in O–H and N–H stretching (carboxylic acid-alcohol and primary amines, respectively) that is constituted in the 3100–3500 cm^{-1} bandwidths. This observation unveiled another compositional difference between the solid bitumen and the in-situ solid bitumen generated from *Tasmanites* telalginite.

CH$_3$/CH$_2$ ratio also varied from 0.50 to 0.87, with a mean value of 0.70 inside the bituminized *Tasmanites*. This ratio for the solid bitumen particle was calculated from 0.04 to 0.99 with a mean value of 0.46 (Table 1.2). This pointed to the presence of long aliphatic chains and a low degree of branching in the solid bitumen particle compared to the bituminized *Tasmanites*, which could have led to the generation of different type of hydrocarbons (e.g., oil vs. gas and condensate, respectively). Relatively higher CH$_3$/CH$_2$ ratio in the bituminized *Tasmanites* could be due to aliphatic chains cracking, which generally results in the increase of the CH$_3$/CH$_2$ ratios [19] during the process of in-situ bituminization of *Tasmanites*.

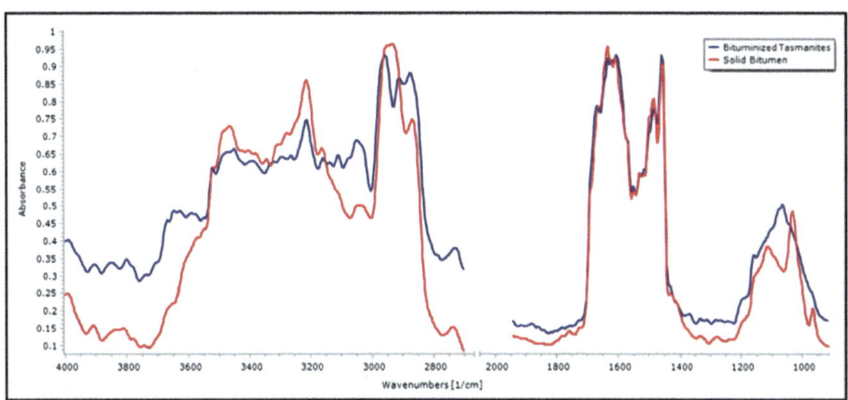

Fig. 1.7 Averaged spectra for solid bitumen and bituminized Tasmanites, indicating functional group disparity amongst two generations of solid bitumens

Table 1.3 Comparison between the aliphatic and aromatic functional groups of the solid bitumen and the bituminized *Tasmanites* from the averaged spectra

Functional group region (cm^{-1}) area	Bituminized *Tasmanites*	Solid bitumen
AL 1370	10.58	12.50
AL 1450	3.92	10.42
AL 1000–1100	43.93	36.56
AR H3000–3100/ AL 2800–3000	0.42	0.36
AR 1490	57.65	22.48
AR 1600	61.42	72.18

They observed that calculating 'A' and 'C' factors for these two bitumen (Table 1.2) revealed strong heterogeneity for both particles on the Ganz and Kalkreuth [13] diagram (Fig. 1.8). While both particles had comparatively similar 'A' factor mean values (0.63 vs. 0.62, in bituminized *Tasmanites* and the solid bitumen, respectively), the solid bitumen was found to have smaller 'C' factor mean value (0.37 vs. 0.47 in bituminized *Tasmanites*). Almost all data points from both particles followed the evolution trend of kerogen Type I and II (Fig. 1.8). Therefore, one could postulate a parent maceral of kerogen Type I or Type II for the solid bitumen. Factor 'A' is commonly used to evaluate petroleum generation potential of source rocks [13]. Both particles exhibited average values of 0.62–0.63 for 'A' factor (Table 1.2) seemingly resulting in similar petroleum generation potential.

Ultimately, the standard deviation of all indices from each particle (Table 1.2) defined a notable chemical heterogeneity in both OM particles, with relatively more homogeneity in the solid bitumen compared to the bituminized *Tasmanites*. Abarghani et al. [3] suggested that it could be due to a full conversion of the parent maceral to solid bitumen compared to the bituminized *Tasmanites* which was still undergoing the process of in-situ bituminization. They confirmed this from the lack of fluorescence emission from the solid bitumen in comparison to the bituminized *Tasmanites*, which exhibited a stronger fluorescence inherited from the parent *Tasmanites* telalginite.

B   Early mature samples

1. Solid Bitumen

On comparing the spectra from ROI of Zone A (Fig. 1.9) of the early mature sample, Abarghani et al. [3] observed noticeable homogeneity compared to the peak mature sample solid bitumen. Generally, more heterogeneity is expected in the early mature sample considering that OM that has yet to endure thermal maturity should contain numerous variations in its chemical composition. However, the solid bitumen's parent maceral intrinsic natural homogeneity (e.g., telalginite derived from Botryococcus or Gleocapsomorpha Prisca) [10] would produce more homogenous solid bitumen in early stages of thermal maturity compared to late mature one. Major discrepancies in the spectra was observed in the 1000–1150 cm^{-1} (S=O and C–O stretching,

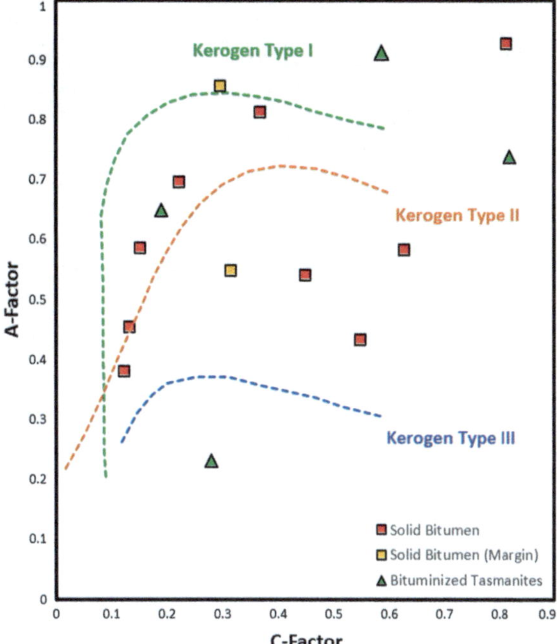

Fig. 1.8 Ganz and Kalkreuth [13] diagram for the peak mature sample's individual macerals. Both generations of solid bitumen are following the evolution trends of kerogen types I and II

sulfoxide, alcohols, aliphatic ethers) [7], 2800–2900 cm^{-1} (C–H stretching, aliphatic CH, CH$_2$, and CH$_3$) [7, 26], and 3200 towards 4000 cm^{-1} bandwidths (including of unsaturated C–H, O–H, N–H stretching, alkene, carboxylic acid-alcohol, and a primary amine) (Fig. 1.10). However, chemical mapping revealed the heterogeneity across the observed area in this organic matter particle (Fig. 1.9b, fixed on 2920 cm^{-1}) suggested a higher degree of chemical variation across the probed area.

They noted that CH$_3$/CH$_2$ ratio exhibited a significant difference between the early mature solid bitumen and the peak mature solid bitumen (0.26 vs. 0.46, respectively). This confirmed the higher potential of petroleum generation for the early mature solid bitumen compared to the peak mature one when considering long aliphatic chains and a low degree of branching in the early mature bitumen. The CH$_3$/CH$_2$ ratio initially demonstrated an increase and then remained relatively unchanged with maturity progression. However, the overall shape of the spectra delineates a relative homogeneity in the early mature solid bitumen where 'A' and 'C' factor (Table 1.4) exhibited significant differences. This observation, along with statistics obtained from other indices especially the AR H3000–3100 cm^{-1}/AL 1370 cm^{-1}, AR C1600 cm^{-1}/AL 1450 cm^{-1}, and AR C1600 cm^{-1}/AL 1370 cm^{-1} confirmed a considerable degree of heterogeneity in the early mature solid bitumen. Ganz and Kalkreuth [13] diagram (Fig. 1.14), suggested a kerogen Type I or I/II as the parent maceral of this bitumen.

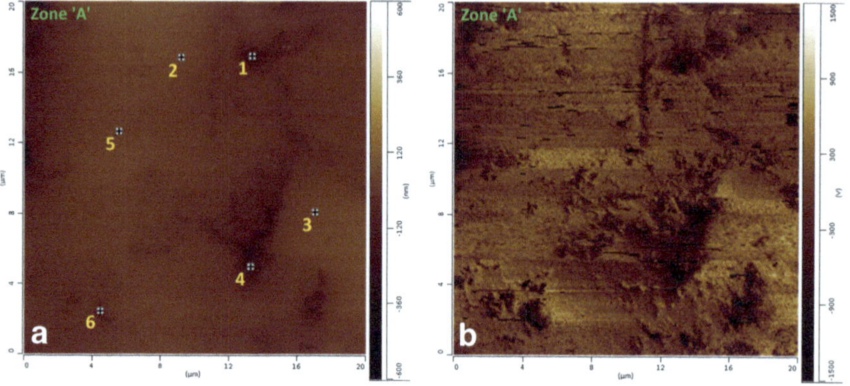

Fig. 1.9 **a** AFM height map (topography) image of zone 'A' showing locations of individual spectral data points labeled 1–6 inside the early mature solid bitumen particle. **b** Same view as in **a** showing IR chemical mapping of aliphatic C–H stretching (2920 cm^{-1}) across zone 'A'

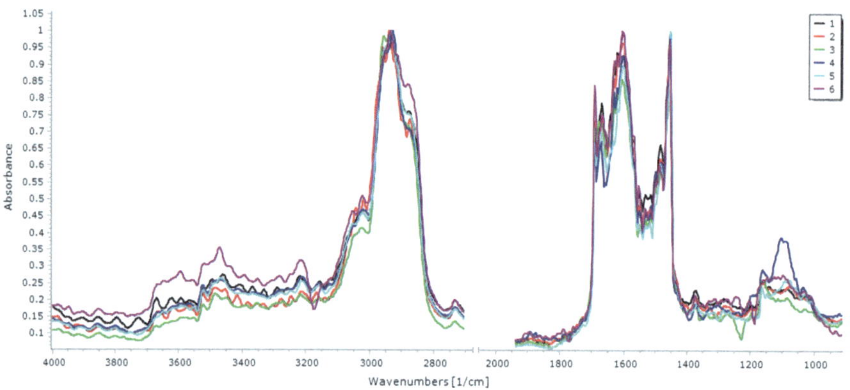

Fig. 1.10 Nano-IR spectra of zone 'A' in the early mature solid bitumen. Numbers correspond to data point locations in Fig. 1.9a

2. Telalginite versus bacterially degraded *Tasmanites*

The interaction of algae with bacteria is widespread in marine waters and is more pronounced during algal bloom periods [24, 27]. Degradation of phytoplankton, particularly during periods of algal bloom, is a complex process that involves two separate steps, cell lysis, and bacterial degradation [27]. Cell lysis results in the degradation of telalginite and in the formation of so-called algal 'ghosts' [27]. This step can be explained as follows: unicellular green algae such as *Tasmanites* and *Leiosphaeridia* incorporate oil globules as flotation [24]. During algal bloom periods, alternating quiescent and turbulent conditions cause the phytoplankton to 'over-float'

Table 1.4 The CH_3/CH_2, Ali/Ox, 'A' and 'C' factors, and ratios of aromatic versus aliphatic nano-IR absorption values for sample 2

| Sample | Spectra | CH_3 | Ali | A_{Factor} | C_{Factor} | $AR_{H3000-3100}$ | $AR_{H3000-3100}$ | $AR_{H3000-3100}$ | AR_{C1600} | AR_{C1600} | AR_{C1600} |
		CH_2	Ox			$AL_{2800-3000}$	AL_{1450}	AL_{1370}	$AL_{2800-3000}$	AL_{1450}	AL_{1370}
Solid bitumen	1	0.69	2.03	0.75	0.29	0.30	1.87	0.75	0.36	2.23	0.89
	2	0.09	2.05	0.15	0.06	0.31	5.86	1.63	0.55	10.58	2.95
	3	0.24	2.15	0.74	0.18	0.27	4.71	30.78	0.55	9.47	61.85
	4	0.43	2.13	0.57	0.15	0.30	2.50	8.89	0.55	4.55	16.14
	5	0.08	2.20	0.74	0.10	0.30	5.31	3.15	0.59	10.59	6.30
	6	0.04	2.25	0.87	0.61	0.31	3.54	9.77	0.09	1.10	3.03
	Average	0.26	2.14	0.64	0.23	0.30	3.97	9.16	0.45	6.42	15.19
	St. Dev	0.25	0.08	0.25	0.20	0.01	1.59	11.24	0.19	4.32	23.49

close to the surface or on the water surface. As a result, phytoplankton is subjected to photo-oxidation of pigments and cells and to severe degradation as a result of intense solar radiation, oxygen oversaturation, and inability to fix nitrogen or photosynthesize [12]. The existence of a stratified water column within the lower and upper Bakken and the presence of photic zone anoxia are supported by the key molecular biomarker gammacerane and by 2,3,6 trimethylaryl- and diaryl isoprenoids [28], respectively, and by the concentration of transitional metals such as Mo, Cr, Ni, and V [4]. According to the above authors, the green sulfur phototrophic bacteria *Chlorobiaceae* that lived in the water column, in the photic zone but below the thermocline of the anoxic (euxinic) zone, acted as primary organic matter producers while anaerobic *Bacterivirous Ciliates* acted as decomposers [4]. Furthermore, degradation of kerogen took place within the stratified water column and led to the formation of bituminite Type III, an amorphous sieve-like, micro-laminated, slightly granular-textured, and non-fluorescing kerogen found within the upper and lower Bakken [1, 4, 27]. Processes such as cell lysis, photo-oxidation, and microbial degradation of algal blooms were very important in the accumulation of high concentrations of organic matter in the Bakken Shale, which occasionally reaches up to 25 wt% TOC. On the other hand, upwelling was not an important factor during the formation of the Bakken Shale [27]. Abarghani et al. [3] examined the telalginite maceral and bacterially degraded Tasmanites to understand whether chemical variation in OM due to a degradation process (Fig. 1.11c) would differ from natural thermal maturation. They noted that IR spectra across zone 'A' on the telalginite body unveiled the highest heterogeneity observed in OM particles examined in their study. This variability in the spectra was observed in almost all bandwidths, particularly between 912 and $1400\,cm^{-1}$ and 2850–$4000\,cm^{-1}$. The IR chemical mapping of the telalginite surface (Fig. 1.11b) fixed on $2920\,cm^{-1}$ also revealed another level of variability in the chemical composition in this particle. This became more special when it was compared to the early mature solid bitumen IR chemical map. This variation in the spectral maxima due to intra-*Tasmanites* chemical heterogeneity has been discussed in other studies conducted using micro-FTIR and elemental analysis [16, 17]. They noted that physico-chemical heterogeneity in the earlier stages of thermal maturity originated from variability in chemical compounds in the OM particles, which was reflected on the IR spectra. This could be attributed to the progression of thermal maturity, which causes less-resistant chemical components to be expelled from the telalginite structure in the upper stages of the oil window. Ultimately, this makes the maceral relatively more homogeneous before it is completely converted into petroleum or other by-products.

They also observed that IR spectra across zone 'B' in the degraded *Tasmanites* suggested a high degree of heterogeneity on the OM surface where the variability in spectra was visible almost in all bandwidths with less variation in their intensity compared to the telalginite particle. The AFM height map (Fig. 1.11c) and the IR chemical map (fixed on $2920\,cm^{-1}$, Fig. 1.11d) explicitly revealed the remnants of the parent *Tasmanites* maceral. Averaged absorption of the spectra from all data

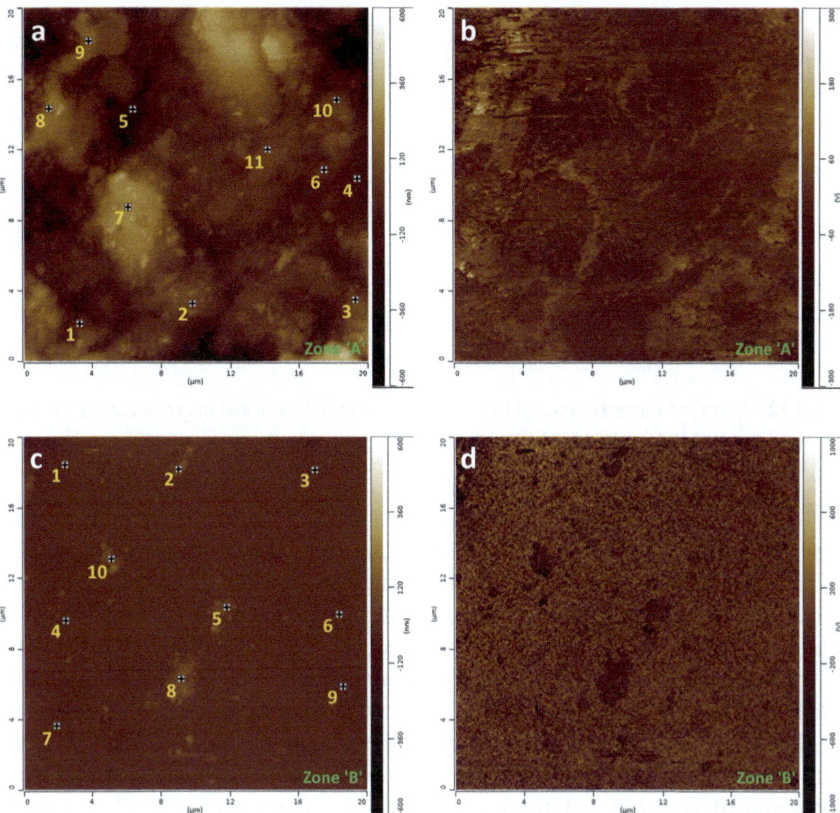

Fig. 1.11 **a** AFM height map (topography) image of zones 'A' of telalginite maceral representing a wide variation in this maceral relief compared to other studied macerals. **b** Same view as in Fig. 1.11a showing IR chemical mapping of aliphatic C–H stretching (2920 cm^{-1}) across zone 'A' and representing a high degree of variation in the surveyed surface's chemical composition. **c** AFM height map (topography) image of zone 'B' of the degraded Tasmanites. The remnants of the parent Tasmanites are exhibiting greater heights compared to the rest of the degraded matrix. **d** Same view as in Fig. 1.11c showing IR chemical mapping of aliphatic C–H stretching (2920 cm^{-1}) across zone 'B' in the degraded Tasmanites. The parent maceral's remnants are clearly recognizable in the chemical map as dark brown spots

points acquired from the parent maceral relicts (Fig. 1.11c: data points 2, 5, 8, and 10) demonstrated very similar characteristics of the telalginite averaged spectra relating these two to one another. The perfect match of the resulting spectra from each individual maceral confirmed the conversion of the parent maceral during the degradation process instead of a thermal maturation where an alteration in the chemical composition of the morphing OM would have been highly anticipated (Fig. 1.12).

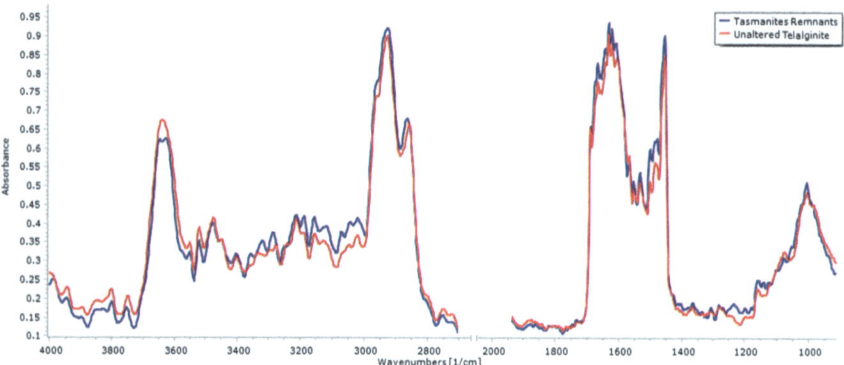

Fig. 1.12 Averaged spectra extracted from the unaltered telalginite and the remnants of the parent Tasmanites inside the degraded matrix

To better outline the impact of degradation process on chemical compounds and functional group alteration in the parent *Tasmanites* telalginite, Abarghani et al. [3] produced averaged spectra for the telalginite and the degraded *Tasmanites* (Fig. 1.13c). They found that the most significant changes on the parent *Tasmanites* telalginite caused by the degradation process occurred in the following bandwidths of the averaged spectra (Fig. 1.13c): fluctuation in the 920–1100 cm^{-1} (aliphatic ethers, alcohols) [7], and then 1200–1400 cm^{-1} bandwidths (asym. C–O stretch, O–H bend, and ethers) [26] where telalginite exhibited higher peak intensities, also in the 1470–1600 cm^{-1} bandwidth (aromatic ring stretch, carboxyl group) [7] where the degraded *Tasmanites* showed significantly higher intensities in relevant peaks. Furthermore, the 2800–3000 cm^{-1} (aliphatic C–H stretching region) was different in each maceral whereby the degraded *Tasmanites* exhibited slightly higher intensities in the peaks from this bandwidth. Additionally, 3000–3580 cm^{-1} (aromatic C–H stretching, and other functional groups including unsaturated C–H, O–H, N–H stretching, alkene, carboxylic acid-alcohol, and primary amine) [7], differed remarkably hence the degraded *Tasmanites* represented a continuously higher intensity in the spectra compared to telalginite. Finally, in the 3580–3680 cm^{-1} bandwidth (O–H stretching, alcohol) telalginite showed higher peak intensity.

Naturally, telalginite has a relatively lower CH_3/CH_2 ratio compared to other macerals due to the presence of long and unbranched aliphatic structures in its chemical structure [19, 20]. However, Abarghani et al. [3] in their study found this ratio to be smaller in zone 'B' (bacterial degraded *Tasmanites*) compared to zone 'A' (telalginite). This suggested a higher amount of initial CH_3/CH_2 ratio in the parent *Tasmanites* maceral compared to the telalginite. They also observed dissimilarities in the ratio of AR H3000–3100 cm^{-1}/AL 2800–3000 cm^{-1} (0.37 vs. 0.28 in degraded *Tasmanites* and telalginite, respectively) showing an increase in aromatic compounds of the parent maceral (i.e., *Tasmanites* telalginite) during the degradation process. The 'A' factor interval of the telalginite (Table 1.5) was larger compared to

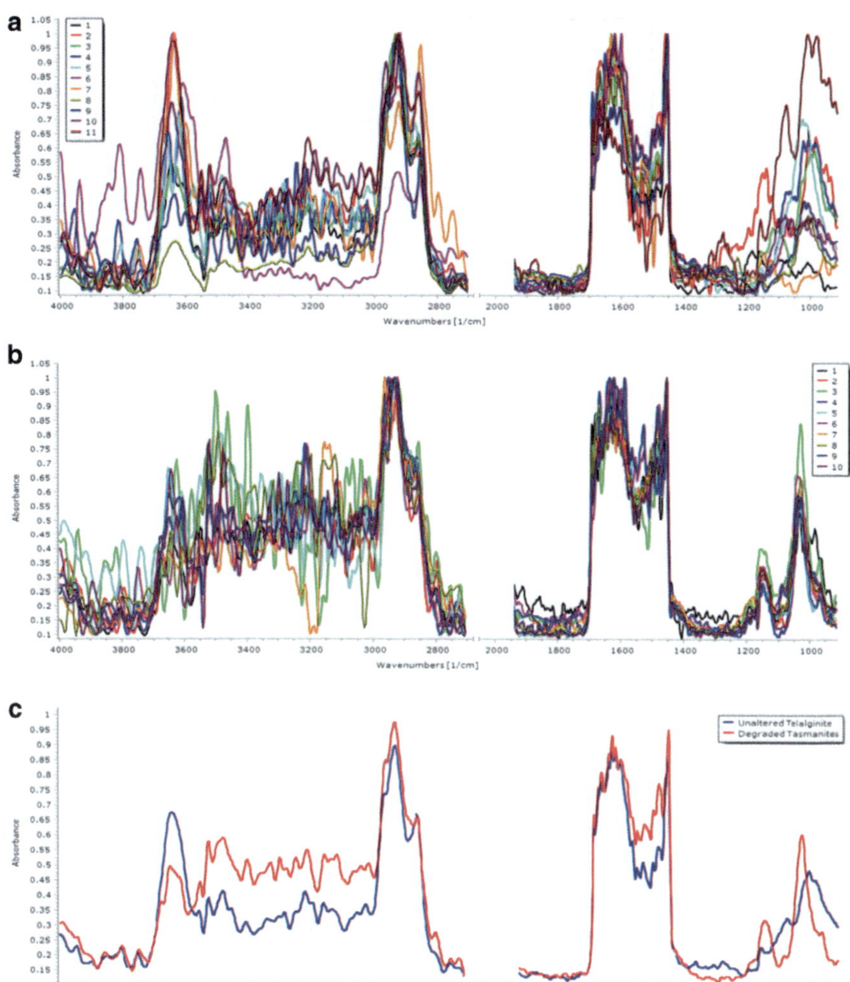

Fig. 1.13 **a** Nano-IR spectra of Zone 'A'; numbers correspond to data point locations in Fig. 1.11a. Considerable heterogeneity is represented by the acquired IR absorption spectra. **b** Nano-IR spectra of Zone 'B'; numbers correspond to data point locations in Fig. 1.11c. **c** Averaged spectra for the unaltered telalginite and the degraded Tasmanites

the degraded *Tasmanites* (Fig. 1.14) and mostly followed the kerogen Type I evolution trend. Moreover, factor 'A' showed fewer variations in the degraded *Tasmanites* (Table 1.5) another verification of the higher level of molecular homogeneity. Ultimately, 'C' factor of the telalginite particle was found to have less variability compared to the degraded *Tasmanites* (St. Dev. of 0.12 vs. 0.20 for telalginite and degraded *Tasmanites*, respectively).

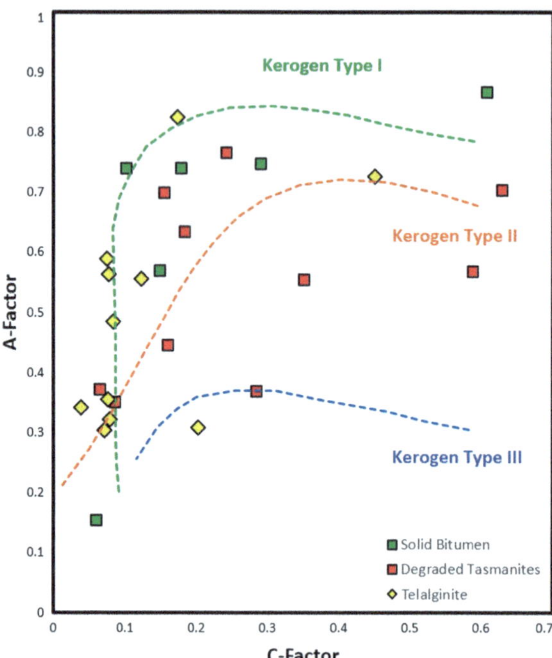

Fig. 1.14 Ganz and Kalkreuth [13] diagram for the early mature samples' individual macerals. Almost all data points are following the evolution trends of kerogen types I and II. Solid bitumen exhibits the widest range in 'A' factor compared to the degraded Tasmanites and telalginite. Wide range of 'C' factor exhibited by degraded Tasmanites confirms the higher degree of heterogeneity in the particle compared to telalginite

They also noted that Except for Ali/Ox ratio, 'A' factor, and AR H3000–3100 cm^{-1}/AL 2800–3000 cm^{-1} indices with a higher St. Dev. values in the telalginite (Table 1.5), all other indices in these zones including 'C' factor, AL H3000–3100 cm^{-1}/AL 1450 cm^{-1}, AR H3000–3100 cm^{-1}/AL 1370 cm^{-1}, AR C1600 cm^{-1}/AL 2800–3000 cm^{-1}, AR C1600 cm^{-1}/AL 1450 cm^{-1}, and AR C1600 cm^{-1}/AL 1450 cm^{-1} denoted higher St. Dev. in the degraded *Tasmanites* particle compared to the telalginite, which reflected a higher degree of heterogeneity. Collectively, their study [3] confirmed that a considerable chemical heterogeneity existed on the surface of the degraded *Tasmanites*, where the degradation process meaningfully enhanced heterogeneity in the final product compared to the parent maceral. They also validated their claim by the Ganz and Kalkreuth [13] diagram where data points from the degraded *Tasmanites* are scattered between kerogen Type I/II to II/III due to the larger 'C' factor interval. Therefore, a higher heterogeneity compared to telalginite could be presumed (Fig. 1.14).

Table 1.5 The CH_3/CH_2, Ali/Ox, 'A' and 'C' factors, and ratios of aromatic versus aliphatic nano-IR absorption values for early mature sample (sample 1)

Sample	Spectra	CH_3 CH_2	Ali Ox	A_{Factor}	C_{Factor}	$AR_{H3000-3100}$ $AL_{2800-3000}$	$AR_{H3000-3100}$ AL_{1450}	$AR_{H3000-3100}$ AL_{1370}	AR_{C1600} $AL_{2800-3000}$	AR_{C1600} AL_{1450}	AR_{C1600} AL_{1370}
Unaltered Telalginite	1	0.74	0.83	0.56	0.12	0.26	6.14	2.25	0.72	16.82	6.16
	2	0.66	0.79	0.82	0.17	0.34	4.61	6.35	0.63	8.60	11.84
	3	0.95	0.81	0.34	0.04	0.28	8.32	2.66	0.56	16.45	5.26
	4	0.48	0.72	0.73	0.45	0.29	4.41	1.46	0.43	6.46	2.13
	5	0.65	0.82	0.56	0.08	0.32	3.66	1.97	0.09	1.00	0.54
	6	0.59	0.51	0.30	0.07	0.17	1.96	0.67	0.12	1.36	0.46
	7	0.41	0.86	0.36	0.08	0.21	5.74	1.80	0.42	11.61	3.63
	8	0.87	0.78	0.59	0.07	0.21	6.97	1.13	0.20	6.38	1.03
	9	0.24	0.88	0.48	0.08	0.23	6.55	1.16	0.15	4.22	0.75
	10	0.15	0.87	0.31	0.20	0.37	12.67	5.07	0.18	6.29	2.52
	11	0.12	1.18	0.32	0.08	0.34	7.40	2.50	0.07	1.43	0.48
	Average	0.53	0.82	0.49	0.13	0.28	6.22	2.46	0.32	7.33	3.16
	St. Dev	0.28	0.16	0.18	0.12	0.06	2.81	1.74	0.24	5.62	3.49
Degraded Tasmanites	1	0.61	0.79	0.35	0.09	0.37	4.72	3.44	0.33	4.11	2.99
	2	0.51	0.78	0.37	0.06	0.33	5.50	2.47	0.52	8.70	3.90
	3	0.20	0.90	0.63	0.18	0.46	17.24	5.61	0.69	25.86	8.42
	4	0.17	0.82	0.77	0.24	0.34	16.73	0.90	0.15	7.19	0.39

(continued)

Table 1.5 (continued)

Sample	Spectra	CH_3 / CH_2	Ali Ox	A_{Factor}	C_{Factor}	$AR_{H3000-3100}$ / $AL_{2800-3000}$	$AR_{H3000-3100}$ / AL_{1450}	$AR_{H3000-3100}$ / AL_{1370}	AR_{C1600} / $AL_{2800-3000}$	AR_{C1600} / AL_{1450}	AR_{C1600} / AL_{1370}
	5	0.49	0.82	0.70	0.16	0.31	8.78	6.08	0.56	15.91	11.02
	6	0.80	0.70	0.55	0.35	0.44	8.82	5.84	0.11	2.29	1.52
	7	0.96	0.87	0.45	0.16	0.39	10.97	2.66	0.09	2.57	0.62
	8	0.45	0.81	0.70	0.63	0.33	7.60	1.77	0.64	14.75	3.44
	9	0.20	0.76	0.57	0.59	0.39	6.45	8.26	0.88	14.62	18.73
	10	0.14	0.86	0.37	0.28	0.37	4.58	2.22	0.69	8.56	4.15
	Average	0.45	0.81	0.55	0.27	0.37	9.14	3.92	0.47	10.46	5.52
	St. Dev	0.28	0.06	0.15	0.20	0.05	4.59	2.37	0.28	7.37	5.72

1.5 Conclusion

Acquired AFM-based NanoIR spectra from specific ROIs on selected organic matters demonstrates significant chemical heterogeneity at the nanoscale. At the same level of thermal maturity in a single sample specimen, two adjacent solid bitumens presented significant chemical variations confirming separate maturity pathways originating from different parent maceral. Comparison of the bacterially degraded *Tasmanites* with the neighboring unaltered telalginite, exposed the role of bacterial degradation in maturity trend of the original *Tasmanites* where during the degradation process, the parent *Tasmanites* had lost its original fluorescence emission. Collectively the process led to an increase in chemical heterogeneity. Chemical heterogeneity is a natural characteristic of organic matter even at the submicron unless the homogeneity is a natural characteristic in some certain type of flora or fauna species.

References

1. Abarghani A, Ostadhassan M, Gentzis T, Carvajal-Ortiz H, Bubach B (2018) Organofacies study of the Bakken source rock in North Dakota, USA, based on organic petrology and geochemistry. Int J Coal Geol 188:79–93
2. Abarghani A, Ostadhassan M, Gentzis T, Carvajal-Ortiz H, Ocubalidet S, Bubach B, Mann M, Hou X (2019) Correlating Rock-Eval™ T_{max} with bitumen reflectance from organic petrology in the Bakken formation. Int J Coal Geol 205:87–104
3. Abarghani A, Gentzis T, Shokouhimehr M, Liu B, Ostadhassan M (2020) Chemical heterogeneity of organic matter at nanoscale by AFM-based IR spectroscopy. Fuel 261:116454
4. Bend SL, Aderoju TE, Wrolson BM, Olajide O (2015) The Bakken formation within the northern part of the Williston Basin: a comprehensive and integrated reassessment of organic matter content, origin, distribution and hydrocarbon potential. In: GeoConvention: New Horizons. Telus Convention Centre, Calgary, AB, Canada, 5pp
5. Binnig G, Rohrer H, Gerber C, Weibel E (1982) Surface studies by scanning tunneling microscopy. Phys Rev Lett 49(1):57
6. Binnig G, Quate CF, Gerber C (1986) Atomic force microscope. Phys Rev Lett 56(9):930
7. Chen Y, Zou C, Mastalerz M, Hu S, Gasaway C, Tao X (2015) Applications of micro-fourier transform infrared spectroscopy (FTIR) in the geological sciences—a review. Int J Mol Sci 16(12):30223–30250
8. Curtis ME, Cardott BJ, Sondergeld CH, Rai CS (2012) Development of organic porosity in the Woodford Shale with increasing thermal maturity. Int J Coal Geol 103:26–31
9. Dazzi A, Prater CB (2017) AFM-IR: technology and applications in nanoscale infrared spectroscopy and chemical imaging. Chem Rev 117(7):5146–5173
10. Engel MH, Macko SA (eds) (2013) Organic geochemistry: principles and applications, vol 11. Springer Science & Business Media
11. Espitalie J, Deroo G, Marquis F (1985) La pyrolyse Rock-Eval et ses applications. Deuxième partie. Rev Inst Fr Petrole 40:755–784
12. Fay P (1983) The blue-greens. Institute of Biology's Studies in Biology 160, Camelot Press, Southampton, 88pp
13. Ganz H, Kalkreuth W (1987) Application of infrared spectroscopy to the classification of kerogentypes and the evaluation of source rock and oil shale potentials. Fuel 66(5):708–711

14. Gentzis T, Carvajal H, Tahoun V, Li C, Ostadhassan M, Xie H, Filho J (2017) A multi-component approach to study the source-rock potential of the Bakken Shale in North Dakota, USA, using organic petrology, Rock-Eval pyrolysis, palynofacies, NMR spectroscopy and LmPy-GCMSMS geochemistry. In: 34th TSOP annual meeting Calgary, Alberta, Canada
15. Goral J (2019) Digital (shale) rock multi-physics: micro-/nano-scale characterization of petrophysical and geomechanical properties of oil-/gas-bearing rocks. Doctoral dissertation, The University of Utah
16. Hackley PC, Kus J (2015) Thermal maturity of Tasmanites microfossils from confocal laser scanning fluorescence microscopy. Fuel 143:343–350
17. Hackley PC, Walters CC, Kelemen SR, Mastalerz M, Lowers HA (2017) Organic petrology and micro-spectroscopy of Tasmanites microfossils: applications to kerogen transformations in the early oil window. Org Geochem 114:23–44
18. Javadpour F, Moravvej Farshi M, Amrein M (2012) Atomic-force microscopy: a new tool for gas-shale characterization. J Can Pet Technol 51(04):236–243
19. Lin R, Ritz GP (1993) Studying individual macerals using i.r. microspectrometry, and implications on oil versus gas/condensate proneness and "low-rank" generation. Org Geochem 20(6):695–706
20. Lis GP, Mastalerz M, Schimmelmann A, Lewan MD, Stankiewicz BA (2005) FTIR absorption indices for thermal maturity in comparison with vitrinite reflectance R0 in type-II kerogens from devonian black shales. Org Geochem 36(11):1533–1552
21. Mastalerz M, Drobniak A, Stankiewicz AB (2018) Origin, properties, and implications of solid bitumen in source-rock reservoirs: a review. Int J Coal Geol 195:14–36
22. Painter PC, Snyder RW, Starsinic M, Coleman MM, Kuehn DW, Davis A (1981) Concerning the application of FT-IR to the study of coal: a critical assessment of band assignments and the application of spectral analysis programs. Appl Spectrosc 35:475–485
23. Peters KE, Cassa MR (1994) Applied source rock geochemistry. In: Magoon LB, Dow WG (eds) The petroleum system: from source to trap, pp 93–120. v. Memoir 60, AAPG
24. Round FE (1981) The ecology of algae. Cambridge University Press, Cambridge, p 653
25. Sanei H, Haeri-Ardakani O, Wood JM, Curtis ME (2015) Effects of nanoporosity and surface imperfections on solid bitumen reflectance (BRo) measurements in sourcerock reservoirs. Int J Coal Geol 138:95–102
26. Solomon PR, Carangelo RM (1988) FT-IR analysis of coal: 2 Aliphatic and aromatic hydrogen concentration. Fuel 67(7):949–959
27. Stasiuk LD (1993) Algal bloom episodes and the formation of bituminite and micrinite in hydrocarbon source rocks: evidence from the Devonian and Mississippian, northern Williston Basin, Canada. Int J Coal Geol 24:195–210
28. Summons RE, Powell TG (1987) Chlorobiaceae in Palaeozoic sea revealed by biological markers, isotopes, and geology. Nature 319:763–765
29. Wang K, Taylor KG, Ma L (2021) Advancing the application of atomic force microscopy (AFM) to the characterization and quantification of geological material properties. Int J Coal Geol 247:103852
30. Webster RL (1984) Petroleum source rocks and stratigraphy of the Bakken formation in North Dakota. In: RMAG guidebook, Williston Basin, Anatomy of a Cratonic Oil Province, pp 268–285
31. Wei L, Wang Y, Mastalerz M (2016) Comparative optical properties of macerals and statistical evaluation of mis-identification of vitrinite and solid bitumen from early mature middle Devonian—lower Mississippian New Albany Shale: implications for thermal maturity assessment. Int J Coal Geol 168:222–236
32. Wood JM, Sanei H, Curtis ME, Clarkson CR (2015) Solid bitumen as a determinant of reservoir quality in an unconventional tight gas siltstone play. Int J Coal Geol 150–151:287–295
33. Yang J, Hatcherian J, Hackley PC, Pomerantz AE (2017) Nanoscale geochemical and geomechanical characterization of organic matter in shale. Nat Commun 8(1):2179

Chapter 2
A Chemo-mechanical Snapshot of In-Situ Conversion of Kerogen to Petroleum

Abstract Organic matter (OM) from various biogenic origins converts to solid bitumen in-situ when it undergoes thermal maturation. It is well documented that during this process, the ratios of both hydrogen and oxygen to carbon will decrease, resulting in an increase in OM aromaticity and molecular chemo-mechanical homogeneity. Although there have been extensive efforts to reveal molecular alteration occurring to OM during conversion, in-situ and continuous observation of such alterations on naturally occurring samples is missing. Therefore, evaluation of previous results cannot be made independent from natural sample variability. In this study, we identified OM particles (*Tasmanites*) that are evolving in-situ into solid bitumen in the Bakken Formation. This in-situ bituminization allows examination of a continuous transformation in OM molecular structure at micron-scale using AFM based IR spectroscopy applied at the transition/interface zone. Moreover, contact mode in the AFM was employed to reveal and relate changes in mechanical properties at a similar scale of measurement. Understanding these chemical and mechanical alterations is important to understand shale reservoir properties and better explain hydrocarbon generation, expulsion, and migration processes at the microscale.

Keywords Kerogen · Organic petrology · AFM-based IR spectroscopy · Chemo-mechanical properties · Solid bitumen reflectance

2.1 Introduction

Organic matter from different biogenic sources, after being deposited along with sediments, during the process of thermal maturation generates hydrocarbons and gets converted to solid bitumen. During the process of maturation or thermal transformation, elemental hydrogen and oxygen are progressively removed, and effectively there is an increase in the content of carbon. Consequently, with progressive thermal maturation, ratios of both hydrogen and oxygen to carbon will decrease, resulting in an increase in OM aromaticity and molecular chemo-mechanical homogeneity. However, depending upon their chemical and molecular structures, different organic

© The Author(s), under exclusive license to Springer Nature Switzerland AG 2024
M. Ostadhassan and B. Hazra, *Advanced Methods in Petroleum Geochemistry*,
SpringerBriefs in Petroleum Geoscience & Engineering,
https://doi.org/10.1007/978-3-031-44405-0_2

matter types behave differently during the progress of thermal maturation [10, 14, 19, 37], resulting in heterogeneity that is still retained in the organic matter at oil and gas window thermal maturity conditions [5, 29, 32, 38]. This is significant since organic matter is abundant in shale plays and influences or controls reservoir mechanical, storage and transport properties [6, 9]. It has been conceived that evaluation of a single component of organic matter in-situ in a continuous manner with progressive thermal maturation will yield critical understandings on the mechanisms of petroleum generation, migration, and accumulation. In fact, some of the researchers have extensively studied conversion of kerogen to petroleum [18, 31, 35]. For example, Hackley et al. [18] studied mid-oil window mature type II kerogen bearing Devonian-Mississippian shale, and inferred that it is possible to identify remnants of the parent kerogen (Tasmanites) along with evidence for its petroleum conversion product in the form of retained solid bitumen. However, the small size of these shale organic components and limitations of typical analytical instruments are the main obstacles to studying them in-situ to reveal chemo-mechanical variations during the conversion process [3, 18]. For example, micro-FTIR cannot examine chemical transitions within individual macerals at sub-micron scale [21, 38]. Abarghani et al. [1], examined the transition/interface zone between remnant oil-prone kerogen and its solid bitumen conversion product is the region of interest (ROI) i.e., where fluorescence emission intensity fades gradually from a parent maceral (in their case Tasmanites) towards its solid bitumen conversion product (the secondary maceral). They selected samples at early and peak mature stages from the Lower Bakken Shale, based on geochemical screening, solid bitumen reflectance and fluorescence properties of the liptinite group macerals. The organic matter present within the samples were examined by AFM-IR spectroscopy at the interface/transition zone between Tasmanites and in-situ produced solid bitumen to map the gradual and continuous chemical and mechanical variations that developed during the conversion of the kerogen to petroleum.

Abarghani et al. [1], using a set of techniques, examined the problems mentioned in the preceding paragraphs. For their study, they selected two samples from the Lower Shale member of the Bakken Formation at early mature and peak mature conditions based on results from programmed pyrolysis (Table 2.1) and ultraviolet (UV) fluorescence characteristics of liptinite group macerals. The results from open-system temperature-programmed pyrolysis-oxidation experiments are presented in Table 2.1. Sample 1 represents a peak mature shale [30], with T_{max} (temperature maxima of the S_2 curve generated from decomposition of kerogen during pyrolysis generating heavier hydrocarbons) of 445 °C, hydrogen index [HI = $(S_2$/TOC) * 100; as a proxy for the atomic H/C ratio of kerogen] of 208 mg HC/g TOC, and production index (PI-Espitalie et al. [12]) of 0.29. Additionally, the dull golden yellow to light-orange fluorescence color of alginite under UV light indicated this sample to be placed in the middle stage of the oil window. Mean solid bitumen reflectance (SBRo) of 0.74% was obtained from 71 measurements in a region of interest (ROI) later examined by AFM-IR spectroscopy (Fig. 2.1a–c). It should be noted that SBRo measurements were used to generate SBRo maps of the ROIs via geostatistical convergent interpolation algorithm method to enable estimation of the equivalent SBRo of the points where AFM-IR spectra were acquired. Sample 2 is early mature as verified by

T_{max} of 437 °C, HI of 534 mg HC/g TOC, and PI of 0.10. An early mature stage is further supported by alginite fluorescence colors of pale greenish-yellow to golden-yellow (Fig. 2.2a–c). Mean solid bitumen reflectance (SBRo) of 0.33% was obtained from 48 measurements in a ROI, later probed via AFM-IR for chemical and mechanical variations. Previous studies have proposed that solid bitumen reflectance can be used to evaluate thermal maturity in the scarcity or absence of vitrinite, especially in marine sediments (such as the Bakken case), and/or Lower Paleozoic strata [15, 20, 22, 25, 26, 33].

Abarghani et al. [1] used polished epoxy-mounted whole-rock pellets for the two Lower Bakken Shale samples and detailed organic petrographic studies were conducted using incident white light and ultraviolet light under oil immersion. As mentioned earlier, they had selected the two shales from different levels of thermal maturity based on liptinite group maceral fluorescence and programmed pyrolysis outcomes. Due to the absence of vitrinite in the samples, they measured solid bitumen reflectance SBRo (%) additionally. Then through interpolation techniques (convergent method), SBRo maps were prepared and used to obtain an SBRo value at the location where IR spectra were acquired. For the purpose of their study, they used a LEICA DM 2500-P with oil immersion objectives and J&M photometer TIDAS S MSP-200 under ASTM protocols [4] with a Sapphire and a GGG (Gadolinium-Gallium-Garnet) calibration standards of 0.589% Ro and 1.716% Ro, respectively, for solid bitumen reflectance (SBRo) and UV light analysis (fluorescence). Analysis under UV light was performed using excitation filter BP 355/425, dichromatic mirror 455, and long-pass filter LP 470 size K.

Results from organic petrography and UV light microscopy suggested that the Tasmanites (unicellular planktonic marine algae) within the samples were transforming to solid bitumen in the ROIs examined by AFM-IR in their study [1], i.e., an in-situ bituminization of kerogen to petroleum occurs. In-situ conversion of alginite to solid bitumen is an important pathway for petroleum generation wherein the alginite is replaced by pre-oil solid bitumen without significant expulsion of hydrocarbons [24, 25]. Hackley et al. [17] also discussed the bituminization of amorphous organic matter including degraded alginite remnants, using nine immature organic-rich source rocks to illustrate the conversion process. In the peak mature sample, fluorescence emission intensity fades while moving inward from the edges to the center of the ROI, suggesting bituminization is more complete towards the center of the Tasmanites. In the ROI, the fluorescence emission from the original alginite is still visible at the margins of the particle (Fig. 2.1b). The organic matter in the ROIs of

Table 2.1 Results from programmed pyrolysis-oxidation source rock analysis for the two samples from the lower shale member of the Bakken formation

Sample ID	Weight	S_1	S_2	T_{max}	S_3	TOC	HI	OI	PI
	mg	mg HC/g	mg HC/g	°C	mg CO$_2$/g	wt%	mg HC/g TOC	mg CO$_2$/g TOC	–
1	60.7	7.46	18.01	445	0.31	8.65	208	4	0.29
2	60.7	10.76	94.24	437	0.35	17.64	534	2	0.10

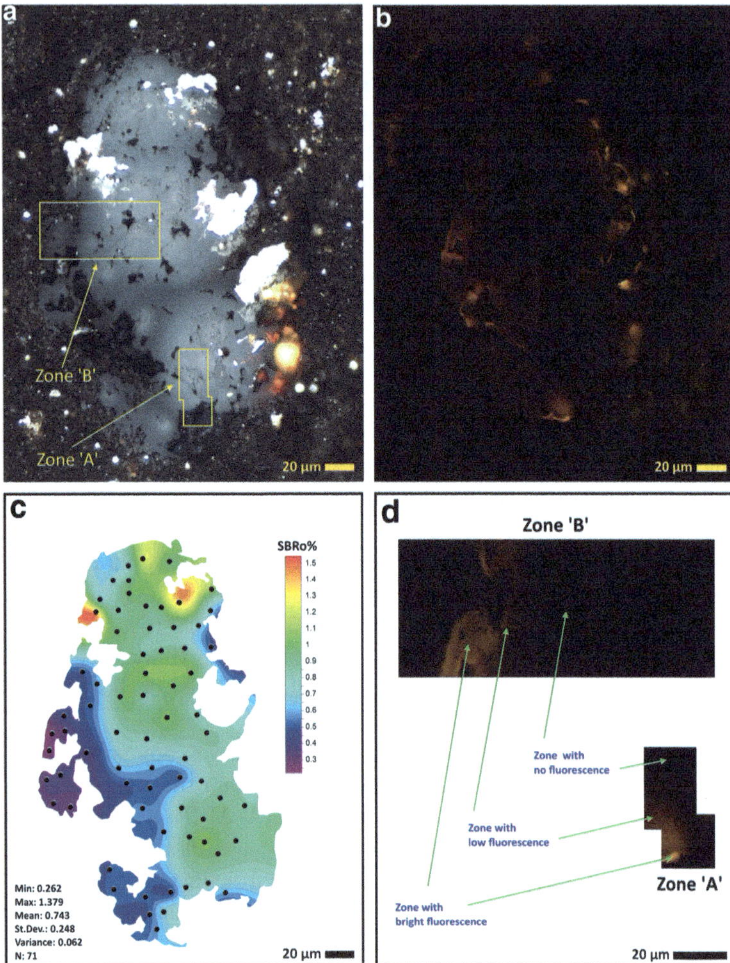

Fig. 2.1 **a** Bedding-orthogonal photomicrograph (50× oil immersion objective under white incident light) of the ROI in peak mature lower Bakken Formation shale (sample 1) showing in-situ bituminization of Tasmanites kerogen to petroleum (solid bitumen). Zones 'A' and 'B' were selected for AFM-IR measurements at the transition/interface zone between fluorescent and non-fluorescent organic matter. **b** Same view as A under UV light. **c** $SBR_o\%$ reflectance map for the ROI based on 71 data points shown as black dots. **d** Higher magnification views of 'A' and 'B' transition zones showing a decrease in fluorescence emission from the margin towards the center of the ROI (boundary lines between subzones with different fluorescence characteristics are approximate). **e** Panoramic height map (topography) image of zone 'B' showing acquisition locations of individual spectral data points (1–7). Red dashed lines are showing the approximate boundaries between fluorescence intensity subzones. **f** Nano-IR spectra of zone 'B'; numbers correspond to locations in **e**. **g** Panoramic height map (topography) images of zone 'A' showing acquisition locations of individual spectral data points (1–9). **h** Nano-IR spectra of zone 'A'; numbers correspond to locations in **g**

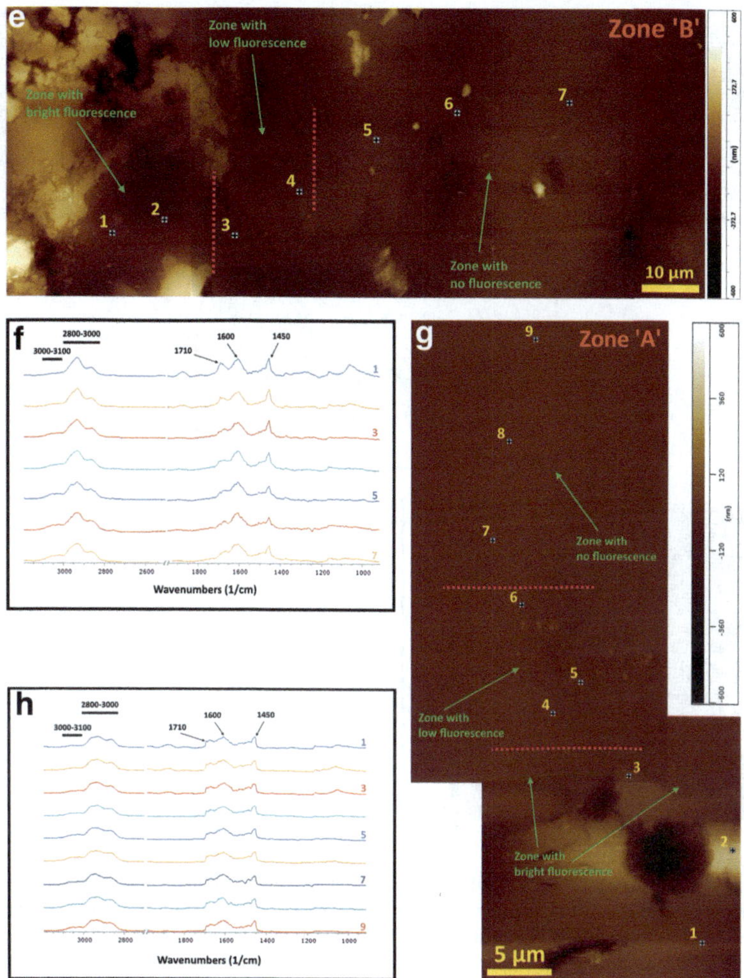

Fig. 2.1 (continued)

both samples displays a remarkable heterogeneity in measured SBRo% (Figs. 2.1c and 2.2c). Therefore, these ROIs present a unique opportunity to investigate the interface or transition zone between the primary alginite kerogen and the secondary petroleum produced in-situ as solid bitumen (Fig. 2.1d).

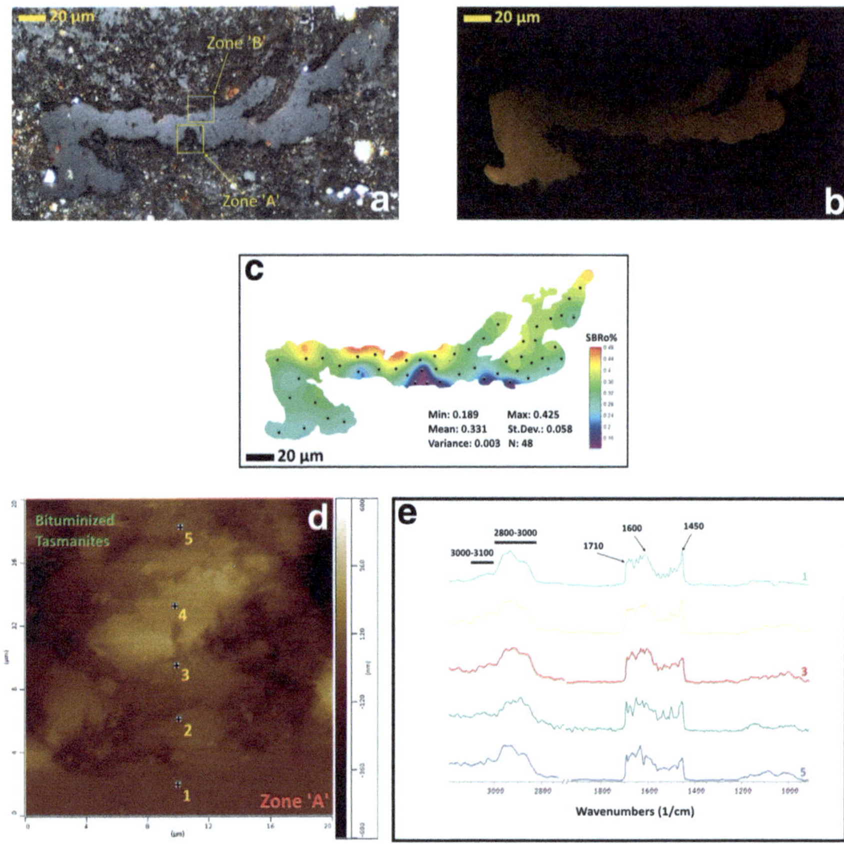

Fig. 2.2 **a** Photomicrograph (50× oil immersion objective under white incident light) of the ROI in early mature lower Bakken Formation shale (sample 2) showing in-situ bituminization of Tasmanites kerogen to petroleum (solid bitumen). Zone 'A' was selected for IR measurements at the transition/interface zone between unaltered and bituminized Tasmanites. **b** Same view as 'A' under UV light showing fluorescence emission ranging from pale greenish-yellow to golden-yellow. **c** $SBR_o\%$ reflectance map for the ROI based on 48 measurements shown as black dots. **d** Height map (topography) image of zone 'A' showing acquisition locations of individual spectral data points labeled 1–5. **e** Nano-IR spectra of zone 'A'; numbers correspond to locations in Fig. 2.2d

2.2   Evolution of Chemical Composition

In their study, Abarghani et al. [1] evaluated geochemical variations across the interface between original alginite kerogen and petroleum (solid bitumen) produced in-situ. Three interface zones were selected based on organic petrography and fluorescence emission intensity, followed by nano-IR spectral acquisition on systematic transects across the transition zones (Figs. 2.1e, g and 2.2d). The following absorption bands were considered for interpretation: (1) 1370 and 1450 cm^{-1}, C–H bending in the methylene group, also including CH_3 and CH_2–CH_3 absorption bands, respectively, (2) 1500–1800 cm^{-1} as the oxygenated group [7, 18] which also includes some important aromatic ring stretch at 1600 cm^{-1} (olefinic/aromatic—[7, 23, 28, 34]), (3) 2800–3000 cm^{-1}, and (4) 3000–3100 cm^{-1} representing aliphatic C–H and aromatic C–H stretching, respectively. Integrated areas under the regions of oxygenated functions (1500–1800 cm^{-1}), aliphatic stretching (2800–3000 cm^{-1}), and aromatic stretching (3000–3100 cm^{-1}) were used to determine IR indices. We also computed the most commonly used indices for chemical variation including CH_3/CH_2 ratio (using 2967 cm^{-1}/2925 cm^{-1}; C–H region, aliphatic stretching—Painter et al. [27]) for aliphatic chain length and branching level [23], 'A' and 'C' factors [13], and also the Ali/Ox ratio [18] defined as the ratio of 2800–3000 cm^{-1} (aliphatic stretch) to 1500–1800 cm^{-1} (oxygenated functions). Considering the CH_3/CH_2 indices, this ratio in the studied samples did not generate any meaningful trends in the data acquisition trajectories on the ROIs from both samples. The average CH_3/CH_2 ratio differs slightly between the two samples (0.26 vs. 0.33 in the peak mature and early mature samples, respectively, here using 2967 cm^{-1}/2925 cm^{-1} ratio, Table 2.2). Moreover, any specific trends were not recognized in both samples regarding the Ali/Ox ratio as well. Natural chemical heterogeneity of the parent *Tasmanites* which has been addressed by other researchers [16] may explain the observations in the CH_3/CH_2 and Ali/Ox ratios in both samples.

Ganz and Kalkreuth [13] defined 'A' (Peak$_{2930}$ + Peak$_{2860}$/Peak$_{2930}$ + Peak$_{2860}$ + Peak$_{1630}$) and 'C' factors (Peak$_{1710}$/Peak$_{1710}$ + Peak$_{1630}$) based on absorption peaks at 2860 and 2930 cm^{-1} (CH_2 and CH_3 aliphatics), 1710 cm^{-1} (carboxyl and carbonyl groups) and 1630 cm^{-1} (aromatic C=C bonds) for quantification of aliphatic and carbonyl/carboxyl group abundances, respectively. They suggested these factors could be used as proxies for H/C and O/C ratios in the traditional Van Krevelen diagram for interpretation of kerogen types and maturity trends [23]. In this study, computation of Ganz and Kalkreuth [13] 'A' and 'C' factors did not result in any meaningful trends along the data acquisition paths in ROIs from the early and peak mature samples (Table 2.2). 'A' factor varies from 0.48 to 0.92 (an average of 0.75) in the early mature ROI and from 0.57 to 0.99 (an average of 0.85) in the peak mature ROI (Fig. 2.3a). 'C' factor ranges from 0.05 to 0.43 (an average of 0.21) in the early mature ROI and from 0.05 to 0.94 (an average of 0.50) in the peak mature ROI. Considering the bivariate plot of Ganz and Kalkreuth [13], it is observed that data from all ROIs follow the evolution paths of kerogen Type I, I/II and possibly II considering the range of the standard error of mean for the calculated 'A' and 'C' factors (Fig. 2.3a). The pseudo Van Krevelen diagram (HI vs. T_{max}) of the bulk samples also confirms the existence of similar kerogen Type II (Fig. 2.3b). Also, Fig. 2.3a indicates the presence of kerogen types I, I/II, and II for the studied samples. However, alginite (here *Tasmanites*) was the only identified maceral in the ROIs of both samples. *Tasmanites* is kerogen Type II, however, the occurrence of the maceral *Tasmanites* as kerogen Type I has been also identified and reported by other researchers [2, 8, 36]. Solid bitumen, amorphous matrix bituminite, granular micrinite, sporinite, marine

Table 2.2 SBR_o and main IR ratios of the two samples studied herein, summarizing transition zones analyzed by nano-IR

	Spectra	SBR_o(%)	CH_3/CH_2	Ali/Ox	'A' factor	'C' factor	ARh 3000–3100/AL 2800–3000	ARh 3000–3100/AL 1450	ARh 3000–3100/AL 1370	ARc 1600/AL 1370
Peak mature sample, zone A	1	–	0.09	0.98	0.86	0.25	0.16	2.43	1.10	3.11
	2	–	0.09	0.96	0.97	0.94	0.19	3.86	1.09	0.38
	3	–	0.96	0.98	0.99	0.80	0.20	1.71	1.28	3.64
	4	0.79	0.10	1.04	0.99	0.72	0.23	1.05	0.60	0.17
	5	0.83	0.07	1.05	0.81	0.10	0.27	3.01	1.88	0.09
	6	0.85	0.04	1.15	0.73	0.16	0.29	6.00	2.78	0.85
	7	0.82	0.02	0.95	0.89	0.79	0.26	2.14	1.01	2.57
	8	0.83	0.03	1.07	0.98	0.79	0.27	4.32	1.58	0.68
	9	0.84	0.08	0.99	0.73	0.05	0.26	3.68	1.05	0.68
Average		0.83	0.16	1.02	0.88	0.51	0.24	3.13	1.37	1.35
St. Dev		0.02	0.30	0.06	0.11	0.36	0.04	1.51	0.64	1.36
Peak mature sample, zone B	1	0.69	0.69	0.84	0.98	0.89	0.16	2.31	3.09	10.48
	2	0.17	0.17	1.00	0.57	0.07	0.17	2.47	10.28	33.14
	3	0.05	0.05	1.01	0.71	0.64	0.22	3.99	5.72	9.63
	4	0.35	0.35	0.92	0.93	0.40	0.22	3.13	8.13	16.19
	5	0.78	0.78	0.93	0.86	0.13	0.27	4.85	11.39	8.41
	6	0.37	0.37	0.86	0.79	0.40	0.28	8.85	4.51	5.83
	7	0.23	0.23	0.81	0.84	0.86	0.28	4.91	0.82	1.44
Average		0.38	0.38	0.91	0.81	0.48	0.23	4.36	6.28	12.16
St. Dev		0.27	0.27	0.08	0.14	0.33	0.05	2.24	3.85	10.28

(continued)

Table 2.2 (continued)

	Spectra	$SBR_o(\%)$	CH_3/CH_2	Ali/Ox	'A' factor	'C' factor	ARh 3000–3100/AL 2800–3000	ARh 3000–3100/AL 1450	ARh 3000–3100/AL 1370	ARc 1600/AL 1370
Early mature sample zone A	1	0.29	0.29	1.04	0.96	0.79	0.25	8.35	2.02	0.21
	2	0.29	0.29	1.03	0.69	0.05	0.39	4.53	25.63	0.74
	3	0.03	0.03	1.01	0.48	0.13	0.33	4.93	15.95	1.61
	4	0.71	0.71	1.01	0.91	0.43	0.37	3.85	4.82	1.91
	5	0.33	0.33	1.17	0.92	0.22	0.39	4.12	71.61	2.61
Average		0.33	0.33	1.05	0.79	0.32	0.35	5.16	24.01	1.42
St. Dev		0.24	0.24	0.07	0.20	0.30	0.06	1.83	28.22	0.95

alginite (Leiosphaeridia, and unidentified prasinophytes or acritarchs), unstructured algal fragments, inertinite, and minor zooclast-like fragments are other constituents of the organic facies which makes the bulk analysis of the samples incapable of accurately characterizing the kerogen type in the pseudo Van Krevelen diagram of Fig. 2.3b.

The ARh 3000–3100 cm^{-1}/AL 2800–3000 cm^{-1} index [23] in the ROIs of the peak mature sample demonstrates an overall increase from areas with higher intensity fluorescence emission to areas without any fluorescence (Fig. 2.1d). This suggests chemical variance and increasing aromaticity between parent kerogen and solid bitumen produced in-situ (Table 2.2). This is correlative to SBR_o increase and decrease in fluorescence emission (Fig. 2.1d). A bivariate plot of solid bitumen reflectance values versus ARh 3000–3100 cm^{-1}/AL 2800–3000 cm^{-1} shows relatively strong linear correlations for zones 'A' and 'B' in the peak mature sample ($R^2 = 0.89$ and 0.91, respectively; Fig. 2.4). SBR_o is controlled by the physicochemical properties of solid bitumen and its interaction with incident light; thus, it is intuitive that SBR_o should have a positive relationship with increasing aromaticity as derived from IR spectra since it reflects the chemical structure of the organic matter. However, no correlation between ARh and SBR_o was observed in zone 'A' of the early mature sample ($R^2 = 0.17$). Since this sample is in the early stages of thermal advance, the lack of correlation between SBR_o and aromatic/aliphatic indices could refer to the natural chemical heterogeneity of the parent *Tasmanites* which has been already addressed by other researchers [16]. As a result of thermal advance, the overall homogeneity of the solid bitumen increases which can improve the correlation between these two

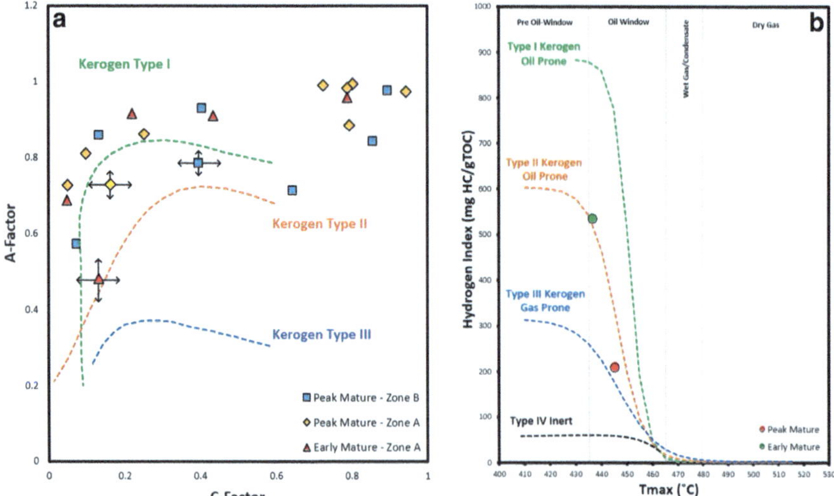

Fig. 2.3 **a** Ganz and Kalkreuth [13] diagram using A- and C-Factor ratios from IR as proxies for H/C and O/C ratios from the traditional Van Krevelen diagram for kerogen typing and maturity trends. **b** Pseudo Van Krevelen diagram (HI vs. T_{max}) for the studied samples

parameters at some point. Though, regardless of the low correlation coefficient that is observed here, an overall increase in SBR_o with increasing aromaticity is clear in the early mature sample as well (Fig. 2.4). An inverse linear relationship between CH_2/CH_3 and SBR_o was reported for the Ohio Shale in both natural and artificially matured (via hydrous pyrolysis) samples [16, 18], however, a similar relationship was not observed in the ROIs of the samples studied herein.

In the early mature sample, the ARc 1600 cm^{-1}/AL 1370 cm^{-1} index exhibits a trend of increasing aromaticity from the margin toward the center of the bituminized *Tasmanites* along the acquired trajectory (data points 2–5). Other indices including CH_3/CH_2, ARh 3000–3100 cm^{-1}/AL 2800–3000 cm^{-1}, ARc 1600 cm^{-1}/AL 1370 cm^{-1}, and Ali/Ox ratios were evaluated but did not produce meaningful trends across the measured transect in *Tasmanites* and its in-situ bituminization residue (Table 2.2). This could be due to the similarity between chemical structures of the parent *Tasmanites* and its bituminization product at the early stages of thermal progression.

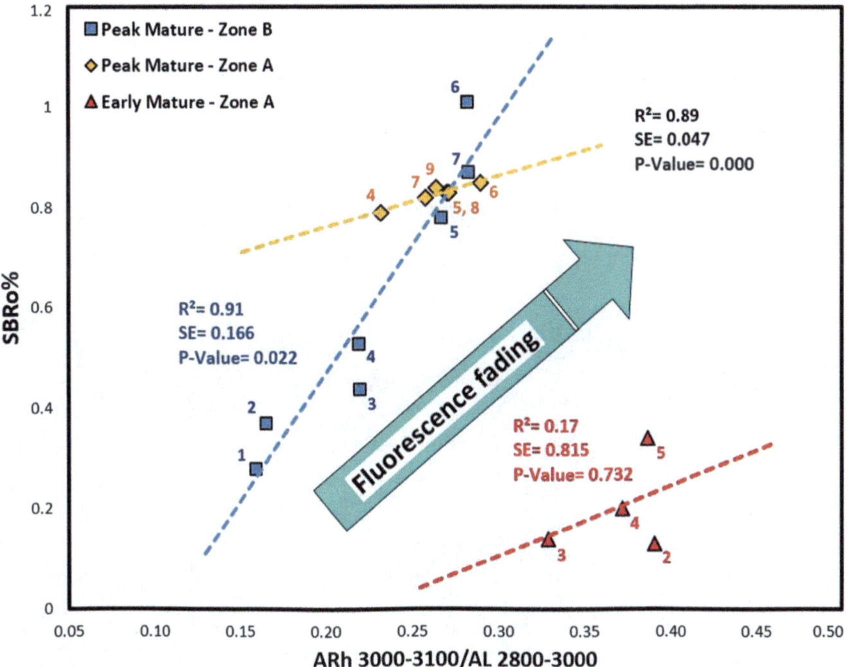

Fig. 2.4 Relationship between ARh 3000–3100/Al 2800–3000 versus SBR_o (%) in zones 'A' and 'B' in sample 1 (peak mature), and zone 'A' of sample 2 (early mature). SE is the standard error. *P*-Value measures the data compatibility with the null hypothesis. Generally, a lower *P*-Value denotes the rejection of the null hypothesis. A combination of high R^2, low SE, and *P*-Value indicates a strong correlation between parameters. Numbers on data points correspond to measurements locations in Fig. 2.1e, and 2.2d

2.3 Nanomechanical Properties

Representative zones between mineral matrix and bituminized *Tasmanites* in the ROI from the early mature sample (zone 'B', Fig. 2.2a), and between unaltered *Tasmanites* and bituminized *Tasmanites* in the ROI of the peak mature sample (zone 'A', Fig. 1a) were selected for mechanical analysis (Fig. 2.6a, b). The objective was to determine the effect of thermal advance in organic matter nanomechanical heterogeneity across the transitions in the ROIs. It was found that Young's modulus in the interface area gradually increased from the original *Tasmanites* kerogen towards its in-situ bituminization residue. This is attributed to thermal maturity advance which causes more labile chemical components to expel from the *Tasmanites* structure, ultimately increasing its stiffness until total conversion to expelled petroleum and residual solid bitumen is achieved. Moreover, we interpret Young's modulus maps to show residual *Tasmanites* kerogen remains as inclusions in bituminized *Tasmanites* (Fig. 2.6b, d) as verified with organic petrography observations (Fig. 2.5). These remnant pieces are more abundant within the organic particle of the early mature sample (Fig. 2.6c), which introduces additional mechanical heterogeneity in this sample compared to the ROI in the peak mature sample. This latest observation can be verified through distribution histograms comparing these two zones for the general range of Young's modulus (in the range of 0–25 GPa—[11]) proposed for organic matter (Fig. 2.6e, f).

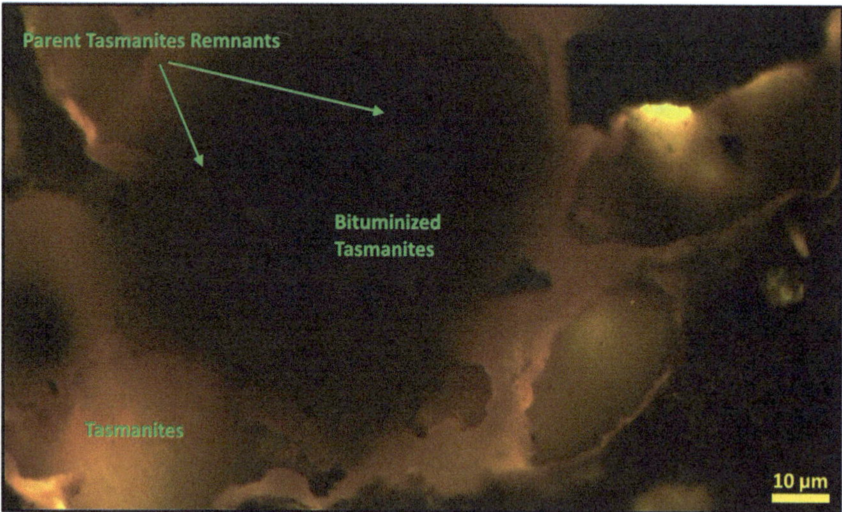

Fig. 2.5 Bituminized Tasmanites from Fig. 2.1b observed at higher magnification showing relicts of fluorescent Tasmanites (parent maceral) internal to its non-fluorescent bituminized residue

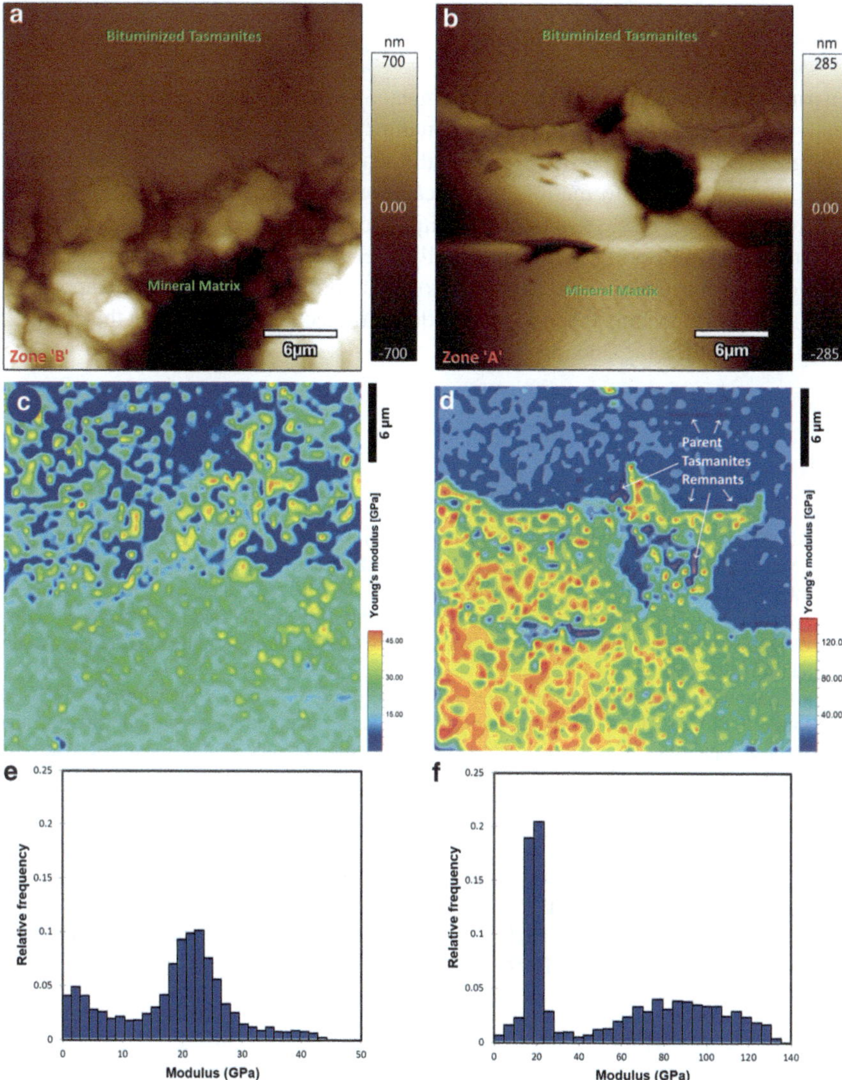

Fig. 2.6 Nanomechanical characterization of the selected interface zone between mineral matrix, unaltered Tasmanites relicts and bituminized Tasmanites in the ROIs from the early and peak mature samples. **a** Height map (topography) of the interface in zone 'B' in the early mature sample. **b** Height map (topography) of the interface in zone 'A' in the peak mature sample. Relict Tasmanites is identified by higher fluorescence emission intensity under UV light. **c** Young's modulus map of image A. **d** Young's modulus map of image **b**. Modulus maps reveals Tasmanites remnants inside bituminized Tasmanites have the same geomechanical properties as unaltered Tasmanites relicts located in the boundary area between the two media. **e** Young's modulus distribution histogram of image **c**. **f** Young's modulus distribution histogram of image **d**

2.4 Conclusion

Thermal maturity can play a role in chemical and mechanical variations in organic matter. IR spectra can represent thermal maturity as well as the heterogeneity within a single organic matter particle, specifically when a single maceral is evolving into solid bitumen. This information can be captured through the relationship between different IR ratios (e.g., aromatic and aliphatic compounds, carbonyl/carboxyl) at various stages of thermal advance. Finally, thermal advance (from early to peak), from original maceral (here Tasmanites) to in-situ produced solid bitumen appears to have impacted mechanical heterogeneity due to molecular alterations in the chemical composition of the organic matter.

References

1. Abarghani, A., Gentzis, T., Shokouhimehr, M., Liu, B., & Ostadhassan, M. (2020). Chemical heterogeneity of organic matter at nanoscale by AFM-based IR spectroscopy. Fuel, 261:116454. https://doi.org/10.1016/j.fuel.2019.116454
2. Aderoju TE (2016) Organic geochemical assessment of the upper and lower members of the Bakken Formation, southern Saskatchewan. Doctoral dissertation, Faculty of Graduate Studies and Research, University of Regina
3. al Sandouk-Lincke NA, Schwarzbauer J, Volk H, Hartkopf-Fröder C, Fuentes D, Young M, Littke R (2013) Alteration of organic material during maturation: a pyrolytic and infrared spectroscopic study of isolated bisaccate pollen and total organic matter (Lower Jurassic, Hils Syncline, Germany). Org Geochem 59:22–36
4. ASTM (2014) D7708-14: standard test method for microscopical determination of the reflectance of vitrinite dispersed in sedimentary rocks
5. Bordenave ML (ed) (1993) Applied petroleum geochemistry vol 524. Technip, Paris
6. Chen J, Xiao X (2014) Evolution of nanoporosity in organic-rich shales during thermal maturation. Fuel 129:173–181
7. Chen Y, Zou C, Mastalerz M, Hu S, Gasaway C, Tao X (2015) Applications of micro-Fourier transform infrared spectroscopy (FTIR) in the geological sciences—a review. Int J Mol Sci 16(12):30223–30250
8. Christopher JE (1961) Transitional Devonian-Mississippian formations of southern Saskatchewan: Department of Mineral Resources 66 [Regina], 103p
9. Curtis ME, Cardott BJ, Sondergeld CH, Rai CS (2012) Development of organic porosity in the Woodford Shale with increasing thermal maturity. Int J Coal Geol 103:26–31
10. Dembicki H (2016) Practical petroleum geochemistry for exploration and production. Elsevier, 342p
11. Eliyahu M, Emmanuel S, Day-Stirrat RJ, Macaulay CI (2015) Mechanical properties of organic matter in shales mapped at the nanometer scale. Mar Pet Geol 59:294–304
12. Espitalie J, Deroo G, Marquis F (1985) La pyrolyse Rock-Eval et ses applications. Deuxième partie. Rev Inst Fr Petrole 40:755–784
13. Ganz H, Kalkreuth W (1987) Application of infrared spectroscopy to the classification of kerogen types and the evaluation of source rock and oil shale potentials. Fuel 66(5):708–711
14. Garcette-Lepecq A, Derenne S, Largeau C, Bouloubassi I, Saliot A (2000) Origin and formation pathways of kerogen-like organic matter in recent sediments off the Danube delta (northwestern Black Sea). Org Geochem 31(12):1663–1683

15. Gentzis T, Goodarzi F (1990) A review of the use of bitumen reflectance in hydrocarbon exploration with examples from Melville Island, Arctic Canada. Rocky Mountain Section (SEPM)

16. Hackley PC, Kus J (2015) Thermal maturity of Tasmanites microfossils from confocal laser scanning fluorescence microscopy. Fuel 143:343–350

17. Hackley PC, Valentine BJ, Hatcherian JJ (2018) On the petrographic distinction of bituminite from solid bitumen in immature to early mature source rocks. Int J Coal Geol 196:232–245

18. Hackley PC, Walters CC, Kelemen SR, Mastalerz M, Lowers HA (2017) Organic petrology and micro-spectroscopy of Tasmanites microfossils: applications to kerogen transformations in the early oil window. Org Geochem 114:23–44

19. Hazra B, Varma AK, Bandopadhyay AK, Mendhe VA, Singh BD, Saxena VK, Samad SK, Mishra DK (2015) Petrographic insights of organic matter conversion of Raniganj basin shales, India. Int J Coal Geol 150:193–209

20. Kelemen SR, Walters CC, Kwiatek PJ, Freund H, Afeworki M, Sansone M, Lamberti WA, Pottorf RJ, Machel HG, Peters KE, Bolin T (2010) Characterization of solid bitumens originating from thermal chemical alteration and thermochemical sulfate reduction. Geochim Cosmochim Acta 74(18):5305–5332

21. Khatibi S, Ostadhassan M, Hackley P, Tuschel D, Abarghani A, Bubach B (2019) Understanding organic matter heterogeneity and maturation rate by Raman spectroscopy. Int J Coal Geol 206:46–64

22. Khorasani GK, Michelsen JK (1993) The thermal evolution of solid bitumens, bitumen reflectance, and kinetic modeling of reflectance: application in petroleum and ore prospecting. Energy Sources 15(2):181–204

23. Lis GP, Mastalerz M, Schimmelmann A, Lewan MD, Stankiewicz BA (2005) FTIR absorption indices for thermal maturity in comparison with vitrinite reflectance R0 in type-II kerogens from Devonian black shales. Org Geochem 36(11):1533–1552

24. Liu B, Schieber J, Mastalerz M (2017) Combined SEM and reflected light petrography of organic matter in the New Albany Shale (Devonian-Mississippian) in the Illinois basin: a perspective on organic pore development with thermal maturation. Int J Coal Geol 184:57–72

25. Mastalerz M, Drobniak A, Stankiewicz AB (2018) Origin, properties, and implications of solid bitumen in source-rock reservoirs: a review. Int J Coal Geol 195:14–36

26. Mukhopadhyay PK (1994) Vitrinite reflectance as maturity parameter: petrographic and molecular characterization and its applications to basin modeling. In: Symposium series, vol 570. American Chemical Society

27. Painter PC, Snyder RW, Starsinic M, Coleman MM, Kuehn DW, Davis A (1981) Concerning the application of FT-IR to the study of coal: a critical assessment of band assignments and the application of spectral analysis programs. Appl Spectrosc 35:475–485

28. Painter PC, Starsinic M, Coleman MM (2012) Determination of functional groups in coal by Fourier transform interferometry. Fourier Transform Infrared Spectrosc 4:169–240

29. Pan C, Geng A, Zhong N, Liu J, Yu L (2009) Kerogen pyrolysis in the presence and absence of water and minerals: amounts and compositions of bitumen and liquid hydrocarbons. Fuel 88:909–919

30. Peters KE, Cassa MR (1994) Applied source rock geochemistry. In: Magoon LB, Dow WG (eds) The petroleum system: from source to trap, v. Memoir 60. AAPG, pp 93–120

31. Robison CR (1997) Hydrocarbon source rock variability within the Austin chalk and Eagle Ford shale (upper cretaceous), East Texas, U.S.A. Int J Coal Geol 34:287–305

32. Schito A, Romano C, Corrado S, Grigo D, Poe B (2017) Diagenetic thermal evolution of organic matter by Raman spectroscopy. Org Geochem 106:57–67

33. Schoenherr J, Littke R, Urai JL, Kukla PA, Rawahi Z (2007) Polyphase thermal evolution in the Infra-Cambrian Ara Group (South Oman Salt Basin) as deduced by maturity of solid reservoir bitumen. Org Geochem 38:1293–1318

34. Solomon PR, Carangelo RM (1988) FTIR analysis of coal: 2 Aliphatic and aromatic hydrogen concentration. Fuel 67(7):949–959

35. Stasiuk LD (1994) Fluorescence properties of Palaeozoic oil-prone alginite in relation to hydrocarbon generation, Williston Basin, Saskatchewan, Canada. Mar Pet Geol 11:219–231
36. Stasiuk LD, Osadetz KG, Potter J (1990) Fluorescence spectral analysis and hydrocarbon exploration: examples from Paleozoic potential source rocks, Saskatchewan. Saskatchewan Geol Soc 10:242–251
37. Tyson R (1995) Abundance of organic matter in sediments: TOC, hydrodynamic equivalence, dilution and flux effects. In: Sedimentary organic matter, pp 81–118
38. Yang J, Hatcherian J, Hackley PC, Pomerantz AE (2017) Nanoscale geochemical and geomechanical characterization of organic matter in shale. Nat Commun 8(1):2179

Chapter 3
Bacterial Versus Thermal Degradation of Algal Matter: Analysis from a Physicochemical Perspective

Abstract Bacteria are ubiquitous in depositional environments, especially where anoxic/euxinic conditions prevail. In such environments, sulfate-reducing bacteria play a critical role to supply sulfur as a biogenic source for H_2S through biomass degradation. In the biodegradation process, chemical and mechanical properties of the organic matter alter. In order to document these variations in-situ, selected samples from a deeply buried mudrock (Bakken Formation), were examined through microscopy analysis. Two separate but adjacent telalginite particles were selected; An unaltered telalginite and a bacterially degraded telalginite, which still contained relics of the parent *Tasmanites*. A combination of AFM-based IR spectroscopy with high-resolution amplitude-frequency modulation was used to evaluate and compare the physicochemical variations across these two particles at the nanoscale. Results indicate that all aromaticity indexes increased for both particles but at a higher rate as a result of bacterial degradation. Furthermore, it was found that bacterial degradation imposes a major mechanical heterogeneity to the organic matter under study, which was detected through phase imaging and modulus mapping captured from submicron to micron-scale level, which exposed the remnants of the parent telalginite.

Keywords Bacterial degradation · Kerogen chemo-mechanical properties · AFM-based IR spectroscopy · High-resolution AFM (AM-FM mode) · Organic petrology

3.1 Introduction

The chemical and mechanical alterations of organic matter (kerogen) deposited in sediment and exposed to biodegradation are less understood and rarely documented [19, 42, 54] compared to thermally degraded alterations. The latter process, which is also known as "thermal maturation" or "thermal advance" in the context of geosciences [13, 41, 49, 54, 56, 61] has been studied extensively due to its implications with respect to hydrocarbon generation. The importance of investigating other processes, such as bacterial degradation, which is solely an independent process and

M. Ostadhassan and B. Hazra, *Advanced Methods in Petroleum Geochemistry*,
SpringerBriefs in Petroleum Geoscience & Engineering,
https://doi.org/10.1007/978-3-031-44405-0_3

"

may vary based on environmental conditions (including temperature and pressure), concerns the role it can play on the evolutionary history of organic matter (Hartkopf-Fröder et al. [19] and references therein). It can also better allude to processes that lead to the expulsion of hydrocarbons from organic matter. In this regard, the presence of framboidal pyrite is a manifestation of bacterial activity, whereby sulfate-reducing bacteria decompose the organic matter and provide ingredients for the precipitation of iron sulfide (framboidal pyrite) in anoxic/euxinic depositional environments [14, 32, 33, 34, 46, 48, 53].

Machel [35] stated that microbial sulfate reduction takes place in a temperature range of 0–80 °C. Beyond 80 °C, active H_2S producing hyperthermophilic microbes are very rare, with the exception of hydrothermal vents [21], which aid in the thermogenic decomposition of organic matter. The organic products of thermogenesis can themselves later become the subject of microbial activities if they are exposed to suitable thermal regimes. This means that utilizing the appropriate analytical techniques, in particular $\delta^{18}O$, $\delta^{13}C$, $\delta^{34}S$ isotopes, elemental and chromatographic analysis of the organic matter, crude oils, and sulfate reduction formed minerals, is a necessary requirement to discriminate thermogenic from biogenic processes [35, 36]. Aderoju and Bend [2] utilized saturate and aromatic biomarkers analysis on the Upper and Lower Bakken Shale samples at the Canadian portion of the Williston Basin and confirmed the existence of various bacteria in the depositional environment of the Bakken Shale. They argued that the presence of anaerobic Bacterivorous Ciliatesa could be verified by a considerable concentration of gammacerane biomarkers. The presence of the green sulfur bacteria Chlorobiaceae was also confirmed by the significant abundancy of mentioned bacteria-derived biomarkers including isorenieratene; 2,3,6-trimethyl aryl- and diaryl-isoprenoids.

Synnott et al. [54] reported a reflectance measurement of biodegraded amorphous organic matter (AOM or bituminite) where a positive increase in bituminite reflectance resulting from bacterial activity (sulfate reduction) was detected. They used a visual abundance of framboidal pyrite in photomicrographs as well as sulfur isotope $\delta^{34}S$ fractionation to confirm microbial activity in samples from the Cenomanian–Turonian Second White Specks Formation of the Upper Colorado Group in south-central Alberta. Since the reflectance of any type of organic material can be discussed in terms of organic photonics, the above authors conjecture that, if chemical and mechanical characteristics of the organic particles in-situ (in the presence of mineral matrix) can be quantified by direct means, biogenic versus thermogenic organic products can then be separated from one another.

Abarghani et al. [1], for the first time in their study clearly differentiated bacterially degraded organic matter and thermally matured organic matter from each other at the nanoscale in-situ through a combination of organic petrology, atomic force microscopy-based infrared spectroscopy (AFM-IR), and amplitude modulation-frequency modulation (AM-FM) AFM microscopy. The investigated Tasmanites telalginite, considered as a Type II kerogen, is composed of algaenen (a resistant biopolymer) and chemical structure consisting mainly of long-chain and less-branched aliphatics, with minor contributions from aromatics [10, 17, 57]. In their study, after preliminary bulk geochemical and organic petrology screening of a large number of samples, suitable aliquots were chosen for further microscopic studies. An early mature shale sample (based on a T_{max} value of 436 °C and a HI of 557 mg HC/g TOC) showing pale greenish-yellow to golden-yellow fluorescence colors of unaltered telalginite was retrieved from the Lower Member of the Bakken Formation. A whole-rock aliquot (2 cm × 2 cm) was mounted in epoxy resin and then polished based on ISO 7404-2 [20] standard. The polished section was examined using incident white and UV light under oil immersion for organic matter (maceral) identification, and reflectance measurements was performed utilizing a LEICA DM 2500-P microscope equipped with a J&M photometer TIDAS S MSP-200 following the TSOP/ICCP protocols [4]. A Sapphire and a YAG (Yttrium-Aluminum-Garnet) calibration standard of 0.589% R_O and 0.904% R_O, respectively, were used for bitumen reflectance (BR_O) measurements. Analysis under UV light was performed using excitation filter BP 355/425 nm, dichromatic mirror 455 nm, and long-pass filter LP 470 nm size K.

After identifying the regions of interest (ROI) on the sample surface based on organic petrography composition and microscopy methods, the ROIs were analyzed by a commercial ANASYS NanoIR2-S instrument (Nanoscale IR on < 20 nm films) for IR acquisition and mapping in the range of 912–1958 cm^{-1} and 2700–4000 cm^{-1} with an interval of 4 cm^{-1} and averaging each data point over 256 pulses. The scan rate frequency of the tip was set at 0.2 Hz with the retrace rate of 0.5 Hz, to scan an area of 20 × 20 μm with spatial resolution < 1 μm. AFM contact mode (ANASYS Instruments, PR-EX-NIR2 model probes) was utilized with the resonance frequency of 13 ± 4 kHz and spring constant of 0.07–0.4 N/m, while the bandpass filter was centered at ~ 180 with 50 kHz window. The spectral acquisition time was around 10–15 min. Prior to the acquisition of each spectral data point, IR laser beam was optimized at 2920 cm^{-1}, and at 1450 cm^{-1} and 1600 cm^{-1} wavenumbers. Several IR spectra were collected within the areas of interest from each sample. The localized IR spectra were analyzed using Spectragryph (V.1.2.10; Menges [38]) software following a workflow including: spectra smoothing using Savitzky-Golay method (Interval: 10; polynomial order: 3), baseline removal following linear method, and then spectra normalizing for eliminating the effect of the thickness of the absorbing material on the IR spectral absolute intensities [59], and then creating an averaged spectrum for each sample (calculated based on spectra mean).

The IR intensity of each band of interest was computed from the area under the curve following deconvolution procedures suggested in the literature [6, 7, 9, 59] using the Fityk (V.0.9.8; Wojdyr [58]) software and applying the Lorentzian curve functions (Levenberg–Marquardt algorithm). The 2800–3000 cm^{-1} bandwidths interval was deconvoluted into discrete bands of: symmetrical CH$_2$ stretching (2850 cm^{-1}), symmetrical CH$_3$ stretching (2859 cm^{-1}), C–H stretching (2890 cm^{-1}), asymmetrical CH$_2$ stretching (2919 cm^{-1}), and asymmetrical CH$_3$ stretching (2958 cm^{-1}). Further, 1500–1800 cm^{-1} bandwidths were used to calculate the Ali/Ox ratio (Aliphatic to oxygenated functional groups; [18]). Although this wavenumber interval includes the C=C aromatic ring stretching bands among others, since negligible aromatic responses were observed in acquired spectra in *Tasmanites* specimens, this interval can be considered an exclusive representation of the oxygenated functions. 1300–1500 cm^{-1} bandwidths were resolved into 1370 and 1450 cm^{-1} (C–H bending, aliphatic compounds) to compute the ratio of 3000–3100 cm^{-1} (representing aromatic C–H stretching) to 1450 and 1370 cm^{-1} as aromaticity II and III indices, respectively (Table 3.1). The Aromaticity Index I was obtained using 2800–3000 cm^{-1} (aliphatic C–H stretching), and 3000–3100 cm^{-1} (aromatic C–H stretching) bandwidths (Table 3.1). Finally, the area under the 2958 and 2919 cm^{-1} peaks provided the CH_3/CH_2 ratio.

High-resolution AM-FM Viscoelastic Mapping Mode was utilized by Abarghani et al. [1] for acquiring nanomechanical parameters using an Asylum Research Oxford Instruments Cypher™ ES Environmental AFM. In this method, a piezoelectric driver excites the cantilever into resonance oscillation and the resulting amplitude probes the topography of the surface area of the sample [5]. In the conventional AFM tapping mode, getting accurate measurements is extremely dependent upon selecting an appropriate cantilever with a suitable spring constant that should be roughly matched to the tip−sample stiffness [12, 15, 16, 23]. However, the AM-FM imaging utilizes standard commercially available cantilevers and are capable of producing similar or improved resolution images compared to a conventional tapping mode to characterize materials across an extremely wide range of modulus using the same cantilever (e.g., ~ 0.1 to ~ 200 GPa, Kocun et al. [23]).

Abarghani et al. [1] selected an Au reflective coating, silicon probe with visible apex tip for tapping mode AC160TSA blueDrive AFM cantilever with a resonance frequency of 300 kHz and a spring constant of 26 N/m. The AM-FM mode provides a variety of nanomechanical information including topography, phase, dissipation, modulus, and loss tangent by operating at two separate resonance frequencies of the cantilever, simultaneously. In this mode, two different levels of resonance frequencies were used including a low-frequency mode, which is also known as amplitude modulation (AM) for tapping mode imaging (topography, and loss tangent) and a high-frequency mode, which is known as frequency modulation (FM-stiffness and elasticity, and dissipation). This method enables the measurement of small frequency and phase shifts with very high precision in order to increase the accuracy of the data and reduce uncertainty in measured viscoelastic values significantly. The observed amplitude, phase, and frequency data were input for the quantitative calculation of nanomechanical properties of the surface [23]. By utilizing the AM-FM derived effective spring constant (k_{eff}) and considering the contact mechanics principles,

Table 3.1 Definition of nano-IR structural indices extracted from nano-IR spectra

Indices	Description	Formula
CH_3/CH_2	Aliphatic CH_3/CH_2 ratio	I2958/I2919
Ali/Ox	Aliphatic to oxygenated functions ratio	I2800–3000/I1500–1800
Aromaticity index (I)	Aromatic C–H/aliphatic C–H ratio	I3000–3100/I2800–3000
Aromaticity index (II)	Aromatic C–H/Aliphatic C–H bending ratio	I3000–3100/I1450
Aromaticity index (III)	Aromatic C–H/Aliphatic C–H bending ratio	I3000–3100/I1370

Young's modulus of the material's surface can be quantified. Readers are referred to Labuda et al. [27] and Kocun et al. [23] for the full description of the AM-FM and steps to calculate k_{eff}, and effective Young's modulus (E_{eff}). The effective spring constant k for a homogenous elastic material in contact with a plane punch of R (equivalent radius) can be calculated as follows [23, 26, 28, 50]:

$$k_{eff} = 2RE_{eff} \tag{3.1}$$

$$\frac{1}{E_{eff}} = \frac{\left(1 - v_t^2\right)}{E_t} + \frac{\left(1 - v_s^2\right)}{E_s} \tag{3.2}$$

where E_t, v_t and E_s, v_s are Young's modulus and Poisson's ratios of the tip and sample, respectively. Abarghani et al. [1], used $v = 0.3$ for the organic matter based on Kumar et al. [25] and Eliyahu et al. [12]. In AM-FM mode, calibration is done with materials of known mechanical parameters as a reference, thus geometrical shape of the tip does not need to be accounted for [23, 28].

3.2 Organic Petrology

The presence of unaltered telalginite and bacterially degraded *Tasmanites* telalginite (kerogen Type II) only a few micrometers apart (Fig. 3.1a, b) offers a unique opportunity to compare the physicochemical properties of biodegradation and thermal maturation processes simultaneously. They selected these two separate particles based on subjective characteristics including: their morphologies and internal reflections (the "peacock" colors), slightly grainy texture that appears in patches within the biodegraded telalginite under white incident light, and concentration of framboidal pyrite in the center of the particle as a sign of bacterial activity and degradation of algae [14, 46, 48]. ROI 'B' covers part of the bacterially degraded *Tasmanites* while ROI 'A' is representing the unaltered telalginite, and both were probed with the AFM-IR and AM-FM mode AFM. At first, they did not encounter the parent *Tasmanites* relicts within the biodegraded particle through organic petrography, though further analysis confirmed the presence of the remnants inside the ROI 'B'.

Incident white light reflectance measurement on the maceral surface creates reflectance contrast images [52, 55]. This is done by measuring the intensity of the reflected light from the surface of the studied maceral by utilizing a photometer equipped microscope that is calibrated with specific standards. Organic petrography of the liptinite group is done with UV light to capture the entire image of the organic matter [52]. Reflectance measurements (Mean $BR_O\% = 0.20$, $N = 56$, St.Dev. $= 0.02$) across the surface of the bacterially degraded *Tasmanites* were used to generate the $BR_O\%$ map (utilizing the geostatistical convergent interpolation algorithm) (Fig. 3.1c) of this particle. UV light examination of the bacterially degraded *Tasmanites* particle demonstrated a decrease in fluorescence intensity compared

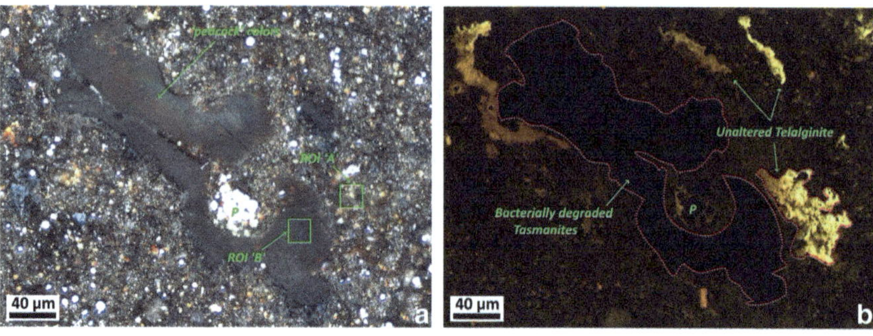

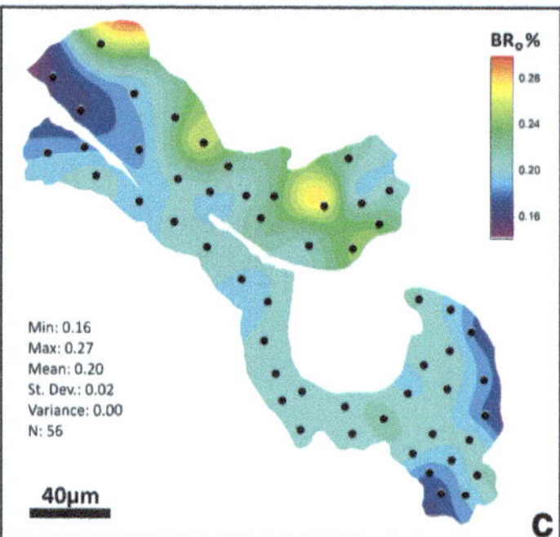

Fig. 3.1 **a** Degraded Tasmanites, recognized based on morphology and texture, under white incident light. Note the concentration of framboidal pyrites (P) in the center of the particle which are surrounded by alginite (under UV light, next figure). **b** Same view as figure **a** under UV light. Unaltered telalginite exhibits pale greenish-yellow to golden-yellow fluorescence colors that suggest early mature stage for this particle. ROI 'A' and 'B' were selected for further IR spectra and AFM measurements. Both photomicrographs were taken using a 50× oil immersion objective. **c** Reflectance map (BR$_O$%) for the bacterially degraded Tasmanites, based on 56 data points shown as black dots that exhibits the chemical heterogeneity across the surface of the particle. Figures **a** and **b** modified from Abarghani et al. [1]

to the adjacent telalginite particles, while the entire bacterially altered *Tasmanites* particle did not exhibit any fluorescence emission.

3.3 Chemical Mapping

Using the AFM-IR technique, Abarghani et al. [1] acquired IR spectra throughout the selected ROIs with the AFM-IR technique. They calculated several spectral indices in order to interpret existing chemical functional groups within the investigated ROIs, including CH$_3$/CH$_2$, Ali/Ox, AR 3000–3100 cm^{-1}/AL 2800–3000 cm^{-1} (Aromaticity Index I), AR 3000–3100 cm^{-1}/AL 1450 cm^{-1} (Aromaticity Index II), and AR 3000–3100 cm^{-1}/AL 1370 cm^{-1} (Aromaticity Index III) (Table 3.2).

Figure 3.2 is the chemical map of the surveyed ROIs (A and B) fixed on 2920 cm^{-1} (aliphatic C–H stretching) showing the IR data points within each ROI, the averaged IR spectrum on each ROI based on the localized IR spectra acquired on the unaltered telalginite in ROI 'A', and the bacterially degraded *Tasmanites* and the parent *Tasmanites* remnants in ROI 'B'. These relicts are recognized based on the significant similarity between the averaged spectra extracted from the unaltered telalginite and the data points no. 2, 5, 8 and 10 within the degraded *Tasmanites* matrix (Fig. 3.2c). This confirms that both particles have the same origin at least based on their chemistry and chemical structure. Furthermore, in the chemical map fixed on 2920 cm^{-1} (Fig. 3.2a) these data points are easily delineated based on their darker color compared to the biodegraded organic matter surrounding them.

Table 3.2 summarizes several major IR indices for these three different organic particles. Lin and Ritz [29] presented CH$_3$/CH$_2$ ratio to characterize the organic matter in geomaterials. Based on this index, a lower CH$_3$/CH$_2$ ratio demonstrates the presence of long aliphatic chains and a low degree of branching, thus pointing to higher oil generation potential. Higher values of CH$_3$/CH$_2$ refer to the shortening of the mean length of aliphatic chains and higher branching in organic material during thermal advance (Table 3.2). Comparison of this ratio among these particles did not show any specific trend, which could be due to the intrinsic difference in the chemical structure of various types of telalginite species and also the result of the natural chemical heterogeneity of *Tasmanites* [17].

Figure 3.2c represents the averaged IR spectrum for each particle including the unaltered telalginite, parent *Tasmanites* remnants, and bacterially degraded *Tasmanites* based on several localized IR spectra within the ROIs. Here, a distinct difference in chemical composition between these particles is observed, which becomes significant in the following bandwidths: 1000–1150 cm^{-1}, 1250–1400 cm^{-1}, 1450–1550 cm^{-1}, 1700–1950 cm^{-1}, 3000–3550 cm^{-1}, 3550–3700 cm^{-1} and where the impact of biodegradation on the parent *Tasmanites* becomes evident. These specific bandwidths correspond to aliphatic ethers, alcohol compounds (1000–1150 cm^{-1}), asymmetrical C–O stretch, O–H bend, and ethers (1250–1400 cm^{-1}), aromatic ring stretch and the carboxyl group (1450–1550 cm^{-1}), C=O stretching including esters, ketone, aldehyde, and carboxylic acid (1700–1950 cm^{-1}), aromatic C–H stretching, N–H stretching, unsaturated C–H, O–H, alkene, carboxylic acid-alcohol and primary amine (3000–3550 cm^{-1}), and O–H stretching and alcohol (3550–3700 cm^{-1}) [6, 29, 43, 51]. Additionally, a sharp increase in aromatic compounds in the bandwidths 3000–3100 cm^{-1} (aromatic C–H stretching) is more prominent in the bacterially degraded *Tasmanites* compared to the unaltered telalginite (Fig. 3.2c).

Hackley et al. [18] defined the Ali/Ox index for organic matter characterization as the ratio of the aliphatic stretch (2800–3000 cm^{-1}) to the oxygenated functions (1500–1800 cm^{-1}). This index (Fig. 3.2d) exhibits the same mean values when comparing the unaltered telalginite and the relicts of the *Tasmanites* within the biodegraded particle, and decreases slightly in the degraded portion (ROI 'B'). Although this ratio does not exhibit a significant discrepancy between the degraded and unaltered particles for our dataset, the effect of bacterial degradation is confirmed by larger spectral values for the degraded portion of the maceral in the bandwidths of

Table 3.2 The CH_3/CH_2, Ali/Ox, and three indices of aromatic versus aliphatic nano-IR absorption values for the studied sample

ROI	Maceral	Spectra	$\frac{CH_3}{CH_2}$	$\frac{Ali}{Ox}$	$\frac{AR_{3000-3100}}{AL_{2800-3000}}$	$\frac{AR_{3000-3100}}{AL_{1450}}$	$\frac{AR_{3000-3100}}{AL_{1370}}$
B	Degraded *Tasmanites*	1	0.61	0.79	0.37	4.72	3.44
		3	0.2	0.9	0.46	17.24	5.61
		4	0.17	0.82	0.34	16.73	0.9
		6	0.8	0.7	0.44	8.82	5.84
		7	0.96	0.87	0.39	10.97	2.66
		9	0.2	0.76	0.39	6.45	8.26
		Average	0.49	0.81	0.40	10.82	4.45
		St. Dev	0.35	0.07	0.04	5.22	2.63
	Tasmanites Rem	2	0.51	0.78	0.33	5.5	2.47
		5	0.49	0.82	0.31	8.78	6.08
		8	0.45	0.81	0.33	7.6	1.77
		10	0.14	0.86	0.37	4.58	2.22
		Average	0.40	0.82	0.34	6.62	3.14
		St. Dev	0.17	0.03	0.03	1.92	1.98
A	Unaltered telalginite	1	0.74	0.83	0.26	6.14	2.25
		2	0.66	0.79	0.34	4.61	6.35
		3	0.95	0.81	0.28	8.32	2.66
		4	0.48	0.72	0.29	4.41	1.46
		5	0.65	0.82	0.32	3.66	1.97
		6	0.59	0.51	0.17	1.96	0.67
		7	0.41	0.86	0.21	5.74	1.8
		8	0.87	0.78	0.21	6.97	1.13
		9	0.24	0.88	0.23	6.55	1.16
		10	0.15	0.87	0.37	12.67	5.07
		11	0.12	1.18	0.34	7.4	2.5
		Average	0.53	0.82	0.28	6.22	2.46
		St. Dev	0.28	0.16	0.06	2.81	1.74

Data retrieved in part from Abarghani et al. [1]

$1500-1800$ cm^{-1} (Fig. 3.2c). Therfore one can deduce that bacterial degradation would increase the oxygenated functions, thus slightly decrease the Ali/Ox ratio.

Calculating three different aromaticity indices (Table 3.1) shows meaningful results and a logical change from the unaltered particle towards the degraded one, with a decrease in intensities of the aliphatic C–H bands and an increase of the aromatic C=C and C–H bands (Fig. 3.2d). Furthermore, results indicate that when comparing bacterially degraded organic matter (here *Tasmanites*) with the thermally matured *Tasmanites*, the AR 3000–3100 cm^{-1}/AL 2800–3000 cm^{-1} index [31] varies in a similar manner. Such a change during thermal maturation has been reported by several authors (e.g., Lin and Ritz [29]; Kruge [24]; Lis et al. [31]; Craddock et al. [7]; Yang [60]; Craddock et al. [8]). This observation can be confirmed with other aromaticity indexes including AR 3000–3100 cm^{-1}/AL 1450 cm^{-1} and AR 3000–3100 cm^{-1}/AL 1370 cm^{-1} (Aromaticity Indices II and III, respectively; Table 3.1) where both indices show increasing values from the unaltered particle towards the degraded one (Fig. 3.2d). Finally, the analysis of functional groups depicts similar

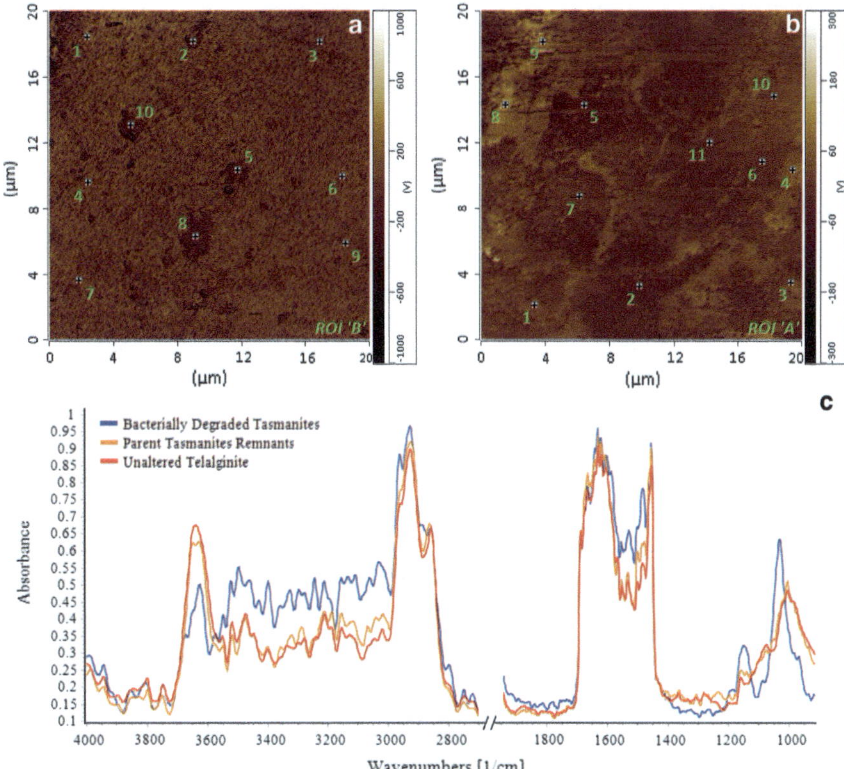

Fig. 3.2 **a** IR chemical map of aliphatic C–H stretching fixed on 2920 cm^{-1} across the ROI 'B', numbers correspond to locations of IR spectra acquiring points. Data points no. 2, 5, 8 and 10 are representing the parent Tasmanites' relicts (dark brown areas) within the biodegraded Tasmanites matrix. **b** IR chemical map of aliphatic C–H stretching fixed on 2920 cm^{-1} across the ROI 'A' and the location of data points. **c** Averaged spectra extracted from the unaltered telalginite, the remnants of the parent Tasmanites inside the degraded matrix, and bacterially degraded Tasmanites. There is a significant similarity between the averaged spectra extracted from the unaltered telalginite (ROI 'A') and the data points no. 2, 5, 8 and 10 within the degraded Tasmanites matrix (ROI 'B'). **d** Histograms for different functional indices and aromaticity indices. Figures **a** and **b** modified from Abarghani et al. [1]

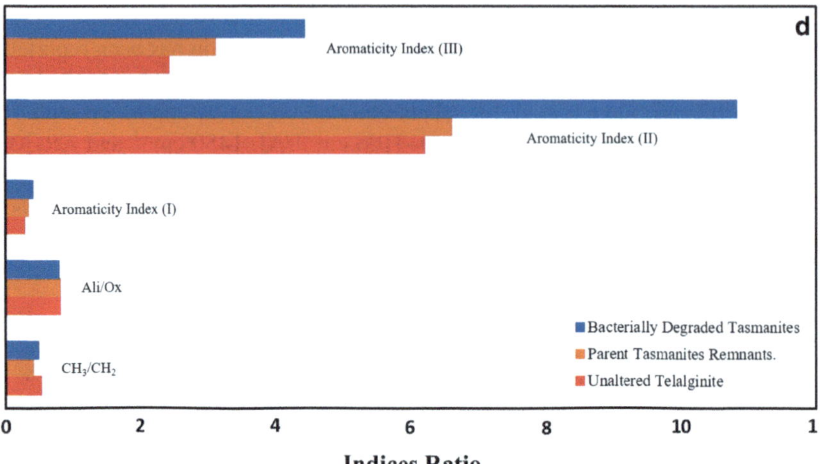

Fig. 3.2 (continued)

variations in aromaticity comparing bacterial degradation and thermal advance where an overall increase is expected.

3.4 Nanomechanical Mapping

The AM-FM mode in AFM is capable of quantifying mechanical properties of the surface at nanometer resolution in a large dynamic range and a fast scan rate [23]. Several magnifications including 20×20 µm, 10×10 µm, and 5×5 µm areas within the ROIs were surveyed for better representation and resolution (Figs. 3.3 and 3.4). In phase imaging (Figs. 3.3 and 3.4d, f), the phase shift between the cantilever driver oscillation signal and the output response signal is monitored simultaneously. This phase lag, which is shown in brighter (low shift) and darker colors (high shift), presents detailed contrast images compared to the heightmap that can better separate different materials based on their physical properties. Phase imaging is very sensitive to the variations in the surface properties and should complement chemical mapping [5, 11, 37]. The ratio of dissipated energy to the stored energy in response to periodic excitation and deformation cycle, known as the viscoelastic loss tangent (tan δ; Mohr et al. [39]; Robertson and Rackaitis [47]; Proksch et al. [44]), defines material mechanical properties. In AM-FM Mode the loss tangent of the sample is measured based on the ratio of dissipated to stored energy during the tip-sample interactions [45]. The loss tangent maps of the surveyed ROIs are exhibited in Figs. 3.3 and 3.4g–i. The hysteresis of the loading and unloading curves (not presented here) during each probing cycle illustrates the energy dissipation and is shown in maps [30] (Figs. 3.3 and 3.4j–l). An increase in tip indentation causes a slight rise in the dissipation curve, which is associated with an increase in the friction force in the contact area [3, 22]. The modulus data in both ROIs (Figs. 3.3 and 3.4m–o) is found to be less than 20 GPa.

Topography maps of ROI 'A' (Fig. 3.3a–c) demonstrate less relief in the organic matter compared to the surrounding mineral matrix while this is opposite in ROI 'B' where the difference in height is significantly smaller (− 60 to 60 nm, compared to − 450 to 450 nm in ROI 'A') comparing the parent maceral (here *Tasmanites*) and the bacterially degraded *Tasmanites*. Phase maps of the ROI 'A' (Figs. 3.3d–f) can easily separate the organic matter from the mineral matrix. Here, the mineral matrix is recognized by darker colors compared to the organic matter, which is shown in green. A relative homogeneity in nanomechanical properties inside the organic matter is denoted by the phase maps in the unaltered telalginite. However, the phase maps for the bacterially degraded *Tasmanites* in ROI 'B' (Fig. 3.4d–f) exposes the parent *Tasmanites* remnants within the surveyed area where the remnants are shown in green and the degraded parts in yellow and yellowish-green. Furthermore, a clear increase in heterogeneity of mechanical properties is discerned from the phase maps (Fig. 3.4d–f). A comparison between two surveyed ROI's phase maps represents the impact of bacterial activity in increasing the mechanical heterogeneity from the parent maceral towards the degraded product.

The viscoelastic properties of the surface [40, 44] were studied in both ROIs, which led to a clear separation between the organic matter and mineral matrix in ROI 'A' (Fig. 3.3g–i) and between parent maceral remnants and bacterially degraded maceral in ROI 'B' (Fig. 3.4g–i). In ROI 'A', the organic matter (unaltered telalginite) shows a relatively strong viscoelastic behavior (tan δ >~ 0.4) compared to the surrounding mineral matrix (tan δ ≪~ 0.4) and a loss tangent in a range of − 0.2 to 1.0 whereas in ROI 'B' the overall loss tangent decreased to the range of − 0.38 to 0.18. It appears that bacterial activity significantly decreased the viscoelastic properties of organic matter. This may be explained by the decomposition of organic matter with the release of sulfur as a source of biogenic H_2S due to bacterial reworking, which could accelerate pore water/rock interactions and increase the volume of authigenic Fe-sulfides with lower viscoelastic properties. Acquired energy dissipation maps in the ROI 'A' (Fig. 3.3j–l) demonstrates a considerable homogeneity across the surveyed area of the unaltered telalginite with a higher average value in comparison to the mineral matrix. This homogeneity is decreased notably in ROI 'B' (Fig. 3.4j–l) due to bacterial activity, exposing the relicts of the parent maceral in the degraded matrix.

The modulus (stiffness) map of ROI 'A' implies a significant mechanical homogeneity in the organic matter regardless of magnification (Fig. 3.3m–o). The homogeneity in stiffness is decreased in ROI 'B' as a result of the bacterial activity (Fig. 3.4m–o). The interesting observation here is that although the parent maceral remnants still exhibit a considerable homogeneity in stiffness, the degraded product in ROI 'B' appears in patches distributed within the parent maceral to impose a considerable heterogeneity. At higher magnifications (Fig. 3.4n–o) this heterogeneity becomes more apparent within the degraded patches as well. These differences in texture originate from separate mechanisms that occur during bacterial degradation

     3   Bacterial Versus Thermal Degradation of Algal Matter: Analysis …

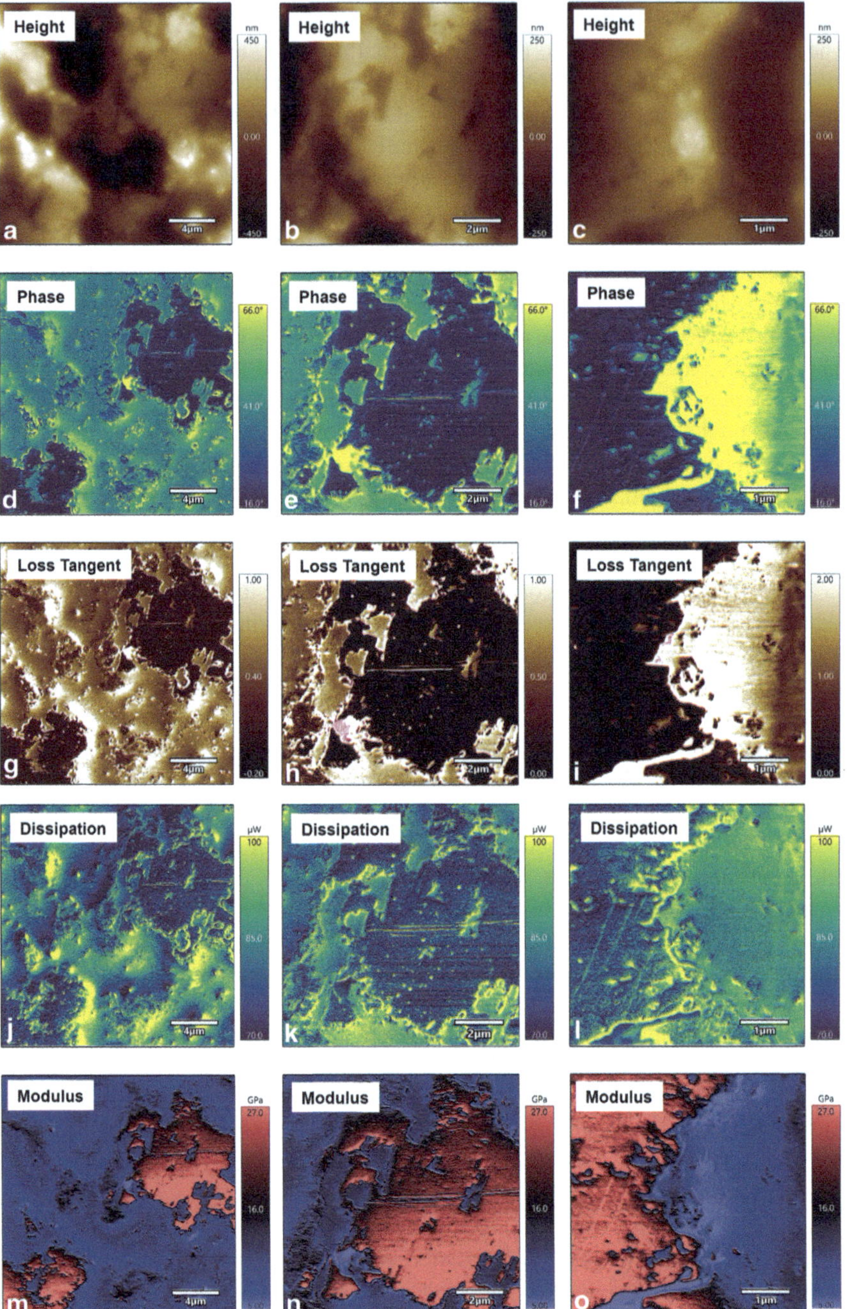

Fig. 3.3 AFM (AM-FM mode) images of the ROI 'A' within the unaltered telalginite with three different resolutions. **a–c** Height maps, **d–f** phase imaging maps, **g–i** loss tangent maps, **j–l** energy dissipation maps, and **m–o** modulus maps

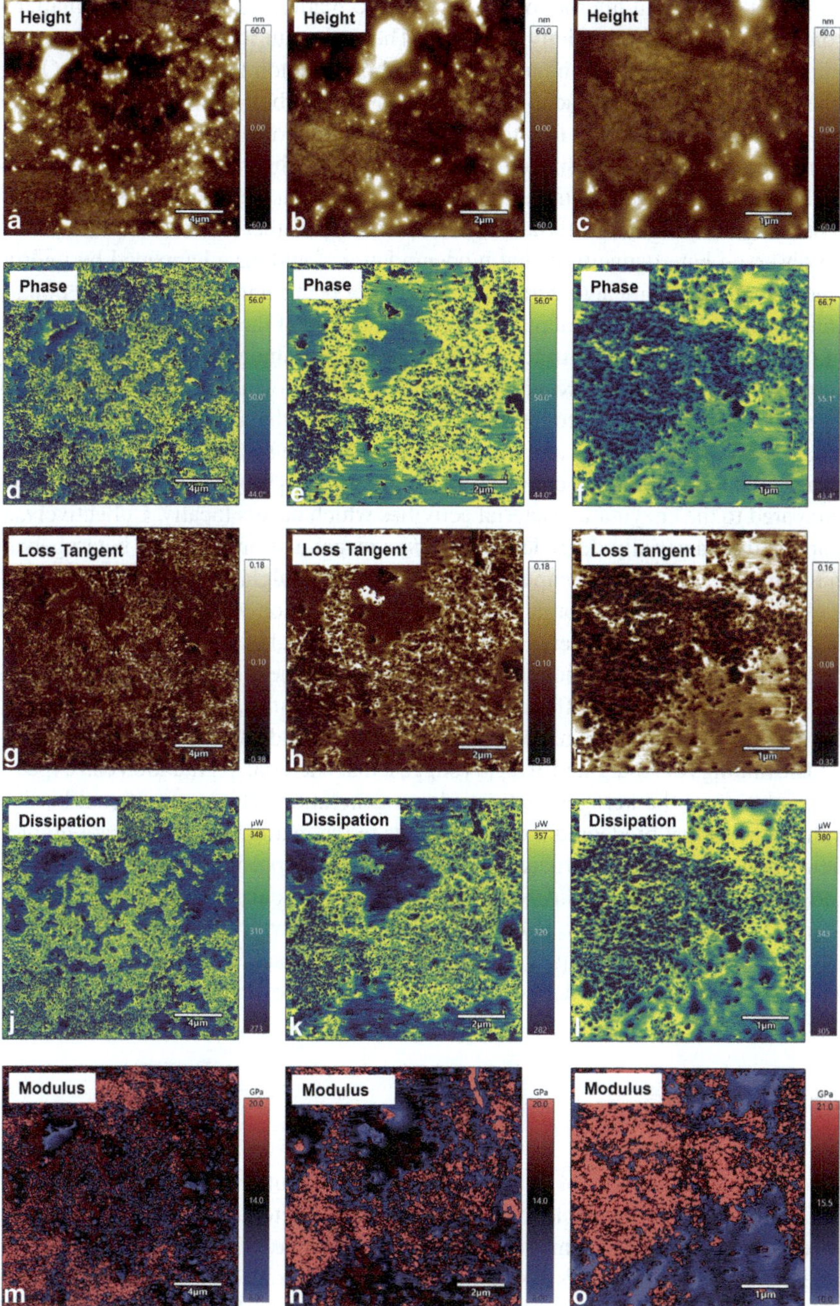

Fig. 3.4 AFM (AM-FM mode) images of the ROI 'B' within the bacterially degraded Tasmanites with three different resolutions. **a–c** Height maps, **d–f** phase imaging maps, **g–i** loss tangent maps, **j–l** energy dissipation maps, and **m–o** modulus maps

compared to thermal alternation of the organic matter. As algal organic matter is exposed to heat, its aromaticity increases. Thermal maturity progression causes the branching to decrease but the main hydrocarbon skeleton to be maintained. On the other hand, during biodegradation, the proteins and carbohydrates that comprise the organic matter, breakdown into their constituent components to provide the energy source for the microogranisms during cell synthesis. This process is selective based on the specificity of the bacterial enzymes, meaning that the site of excision can vary depending on different bacterial enzymes present. Another possible explanation for the observed heterogeneity is that biodegradation could have happened by quorum sensing, where significant degradation might not occur unless the bacterial population density reaches a critical mass. Hence, the nature of the process is not organized and can lead to a patchy surface texture that is not homogenous. However, during thermal maturation which results in production of petroleum and other byproducts, the entire mass is exposed to thermal energy meaning that bonds break in a specific order based on the kinetics and thermodynamics of the system to enforce the phase change. In this evolutionary pathway, the entire system is subjected to heat universally compared to the enzymatic bacterial activities which occurs locally. Collectively, all maps in the bacterially degraded *Tasmanites* reveal a strong level of heterogeneity in mechanical properties compared to the unaltered telalginite. This study presented evolution pathways of the organic matter (biodegradation vs. thermal maturity) that can be separated through detailed physicochemical analysis. Previous studies (e.g., Zeszotarski et al. [61]; Okiongbo et al. [41]; Emmanuel et al. [13]) discussed that organic matter becomes stiffer at higher thermal maturities, and it was realized here that biodegradation would have similar effects independent of temperature.

The findings of Abarghani et al. [1] suggest that bacterial degradation can expedite the evolution of the kerogen and petroleum generation. This is significant because if the impacts that biodegradation could have in the maturation process are ignored, estimations pertaining hydrocarbon generation and expulsion in the basin analysis and petroleum system studies will be erroneous, particularly in depositional environments were algal bloom is confirmed. Therefore, with the availability of more advanced analytical techniques, it is beneficial to precisely characterize the organic matter to more accurately delineate the kerogen (OM) evolution trend.

3.5 Conclusion

Through a combination of AFM-IR and high-resolution AM-FM mode AFM, Abarghani et al. [1] were able to observe the role of bacterial degradation on physicochemical attributes of biomass in-situ. They were able to recognize and reveal the parent *Tasmanites* remnants inside the bacterially degraded particle and show that all aromaticity indices increase, albeit at a different rate, from the unaltered telalginite toward the bacterially degraded *Tasmanites*. The viscoelastic mode of AFM captured that bacterial degradation has an important role in imposing mechanical heterogeneity on the remaining algal matter even at the early stages of maturation.

Finally, bacterial degradation can accelerate the kerogen evolution process, and the production and expulsion of hydrocarbons. This is significant since ignoring this phenomenon can result in considerable errors in petroleum system evaluation and further predictions of proper timing in such studies.

References

1. Abarghani A, Gentzis T, Liu B, Hohlbauch S, Griffin D, Bubach B, Shokouhimehr M, Ostadhassan M (20200 Bacterial versus thermal degradation of algal matter: analysis from a physicochemical perspective. Int J Coal Geol 223:103465
2. Aderoju T, Bend SL (2014) Organic matter variations within the Bakken Shales of Saskatchewan: with implications upon origin and timing of hydrocarbon generation. AAPG Search Discov Article 41333
3. Anczykowski B, Gotsmann B, Fuchs H, Cleveland JP, Elings VB (1999) How to measure energy dissipation in dynamic mode atomic force microscopy. Appl Surf Sci 140(3–4):376–382
4. ASTM (2014) D7708-14: standard test method for microscopical determination of the reflectance of vitrinite dispersed in sedimentary rocks
5. Boussu K, Van der Bruggen B, Volodin A, Snauwaert J, Van Haesendonck C, Vandecasteele C (2005) Roughness and hydrophobicity studies of nanofiltration membranes using different modes of AFM. J Colloid Interface Sci 286(2):632–638
6. Chen Y, Zou C, Mastalerz M, Hu S, Gassaway C, Tao X (2015) Applications of micro-fourier transform infrared spectroscopy (FTIR) in the geological sciences—a review. Int J Mol Sci 16(12):30223–30250
7. Craddock PR, Le Doan TV, Bake K, Polyakov M, Charsky AM, Pomerantz AE (2015) Evolution of kerogen and bitumen during thermal maturation via semi-open pyrolysis investigated by infrared spectroscopy. Energy Fuels 29(4):2197–2210
8. Craddock PR, Prange MD, Pomerantz AE (2017) Kerogen thermal maturity and content of organic-rich mudrocks determined using stochastic linear regression models applied to diffuse reflectance IR Fourier transform spectroscopy (DRIFTS). Org Geochem 110:122–133
9. Craddock PR, Bake KD, Pomerantz AE (2018) Chemical, molecular, and microstructural evolution of kerogen during thermal maturation: case study from the Woodford Shale of Oklahoma. Energy Fuels, 32(4):4859–4872.
10. Dutta S, Greenwood PF, Brocke R, Schaefer RG, Mann U (2006) New insights into the relationship between Tasmanites and tricyclic terpenoids. Org Geochem 37(1):117–127
11. Eaton P, West P (2010) Atomic force microscopy. Oxford University Press
12. Eliyahu M, Emmanuel S, Day-Stirrat RJ, Macaulay CI (2015) Mechanical properties of organic matter in shales mapped at the nanometer scale. Mar Pet Geol 59:294–304
13. Emmanuel S, Eliyahu M, Day-Stirrat RJ, Hofmann R, Macaulay CI (2016) Impact of thermal maturation on nano-scale elastic properties of organic matter in shales. Mar Pet Geol 70:175–184
14. Folk RL (2005) Nannobacteria and the formation of framboidal pyrite: textural evidence. J Earth Syst Sci 114(3):369–374
15. Gates RS, Reitsma MG (2007) Precise atomic force microscope cantilever spring constant calibration using a reference cantilever array. Rev Sci Instrum 78(8):086101
16. Gibson CT, Weeks BL, Abell C, Rayment T, Myhra S (2003) Calibration of AFM cantilever spring constants. Ultramicroscopy 97(1–4):113–118
17. Hackley PC, Kus J (2015) Thermal maturity of Tasmanites microfossils from confocal laser scanning fluorescence microscopy. Fuel 143:343–350
18. Hackley PC, Walters CC, Kelemen SR, Mastalerz M, Lowers HA (2017) Organic petrology and micro-spectroscopy of Tasmanites microfossils: applications to kerogen transformations in the early oil window. Org Geochem 114:23–44

19. Hartkopf-Fröder C, Königshof P, Littke R, Schwarzbauer J (2015) Optical thermal maturity parameters and organic geochemical alteration at low grade diagenesis to anchimetamorphism: a review. Int J Coal Geol 150:74–119
20. ISO 7404-2 (2009) Methods for the petrographic analysis of coals, part 2: methods of preparing coal samples
21. Jørgensen BB, Isaksen MF, Jannasch HW (1992) Bacterial sulfate reduction above 100 C in deep-sea hydrothermal vent sediments. Science 258(5089):1756–1757
22. Kámán J (2015) Young's modulus and energy dissipation determination methods by AFM, with particular reference to a chalcogenide thin film. Period Polytech Elec Eng Comp Sci 59(1):18–25
23. Kocun M, Labuda A, Meinhold W, Revenko I, Proksch R (2017) Fast, high resolution, and wide modulus range nanomechanical mapping with bimodal tapping mode. ACS Nano 11(10):10097–10105
24. Kruge MA (2000) Determination of thermal maturity and organic matter type by principal components analysis of the distributions of polycyclic aromatic compounds. Int J Coal Geol 43(1–4):27–51
25. Kumar V, Sondergeld CH, Rai CS (2012, January) Nano to macro mechanical characterization of shale. In: SPE annual technical conference and exhibition. Society of Petroleum Engineers
26. Labuda A, Kocuń M, Meinhold W, Walters D, Proksch R (2016) Generalized Hertz model for bimodal nanomechanical mapping. Beilstein J Nanotechnol 7(1):970–982
27. Labuda A, Grutter P, Miyahara Y, Paul W, Roy-Gobeil A (2017) U.S. Patent No. 9,671,424. Washington, DC: U.S. Patent and Trademark Office.Chicago
28. Li Y, Yu C, Gan Y, Jiang P, Yu J, Ou Y, Zou DF, Huang C, Wang J, Jia T, Luo Q (2018) Mapping the elastic properties of two-dimensional MoS 2 via bimodal atomic force microscopy and finite element simulation. npj Comput Mater 4(1):49
29. Lin R, Ritz GP (1993) Studying individual macerals using i.r. microspectrometry, and implications on oil versus gas/condensate proneness and "low-rank" generation. Org Geochem 20(6):695–706
30. Lineback J, Davidson M (1982) The Williston basin-sediment-starved during the early Mississippian. In: Williston basin symposium
31. Lis GP, Mastalerz M, Schimmelmann A, Lewan MD, Stankiewicz BA (2005) FTIR absorption indices for thermal maturity in comparison with vitrinite reflectance R0 in type-II kerogens from Devonian black shales. Org Geochem 36(11):1533–1552
32. Love LG (1957) Mircro-organisms and the presence of syngenetic pyrite. Quart J Geol Soc 113(1–4):429–440
33. Love LG, Al-Kaisy AT, Brockley H (1984) Mineral and organic material in matrices and coatings of framboidal pyrite from Pennsylvanian sediments, England. J Sediment Res 54(3):869–876
34. Love LG, Murray JW (1963) Biogenic pyrite in recent sediments of Christchurch harbour, England. Am J Sci 261(5):433–448
35. Machel HG (2001) Bacterial and thermochemical sulfate reduction in diagenetic settings—old and new insights. Sed Geol 140(1–2):143–175
36. Machel HG, Krouse HR, Sassen R (1995) Products and distinguishing criteria of bacterial and thermochemical sulfate reduction. Appl Geochem 10(4):373–389
37. Magonov SN, Elings V, Whangbo MH (1997) Phase imaging and stiffness in tapping-mode atomic force microscopy. Surf Sci 375(2–3):L385–L391
38. Menges F (2018) Spectragryph–optical spectroscopy software, version 1.2. 10. http://www.eff emm2.de/spectragryph/
39. Mohr R, Kratz K, Weigel T, Lucka-Gabor M, Moneke M, Lendlein A (2006) Initiation of shape-memory effect by inductive heating of magnetic nanoparticles in thermoplastic polymers. Proc Natl Acad Sci 103(10):3540–3545
40. Nguyen HK, Ito M, Fujinami S, Nakajima K (2014) Viscoelasticity of inhomogeneous polymers characterized by loss tangent measurements using atomic force microscopy. Macromolecules 47(22):7971–7977

41. Okiongbo KS, Aplin AC, Larter SR (2005) Changes in type II kerogen density as a function of maturity: evidence from the Kimmeridge clay formation. Energy Fuels 19(6):2495–2499

42. Pacton M, Gorin G, Vasconcelos C, Gautschi HP, Barbarand J (2010) Structural arrangement of sedimentary organic matter: nanometer-scale spheroids as evidence of a microbial signature in early diagenetic processes. J Sediment Res 80(10):919–932

43. Painter PC, Starsinic M, Coleman MM (2012) Determination of functional groups in coal by Fourier transform interferometry. Fourier Transform Infrared Spectrosc 4:169–240

44. Proksch R, Kocun M, Hurley D, Viani M, Labuda A, Meinhold W, Bemis J (2016) Practical loss tangent imaging with amplitude-modulated atomic force microscopy. J Appl Phys 119(13):134901

45. Proksch R, Yablon DG (2012) Loss tangent imaging: theory and simulations of repulsive-mode tapping atomic force microscopy. Appl Phys Lett 100(7):073106

46. Raiswell R (1982) Pyrite texture, isotopic composition and the availability of iron. Am J Sci 282(8):1244–1263

47. Robertson CG, Rackaitis M (2011) Further consideration of viscoelastic two glass transition behavior of nanoparticle-filled polymers. Macromolecules 44(5):1177–1181

48. Sanei H, Wood JM, Ardakani OH, Clarkson CR, Jiang C (2015) Characterization of organic matter fractions in an unconventional tight gas siltstone reservoir. Int J Coal Geol 150:296–305

49. Seewald JS (2003) Organic–inorganic interactions in petroleum-producing sedimentary basins. Nature 426(6964):327

50. Sneddon IN (1965) The relation between load and penetration in the axisymmetric Boussinesq problem for a punch of arbitrary profile. Int J Eng Sci 3(1):47–57

51. Solomon PR, Carangelo RM (1988) FT-IR analysis of coal: 2. Aliphatic and aromatic hydrogen concentration. Fuel 67(7):949–959

52. Suarez-Ruiz I, Flores D, Mendonça Filho JG, Hackley PC (2012) Review and update of the applications of organic petrology: part 1, geological applications. Int J Coal Geol 99:54–112

53. Suits NS, Wilkin RT (1998) Pyrite formation in the water column and sediments of a meromictic lake. Geology 26(12):1099–1102

54. Synnott DP, Sanei H, Pedersen PK, Dewing K, Ardakani OH (2016) The effect of bacterial degradation on bituminite reflectance. Int J Coal Geol 162:34–38

55. Taylor GH, Teichmuller M, Davis A, Diessel CFK, Littke R, Robert P (1998) Organic petrology. Gebrüder Borntraeger, Berlin, p 704

56. Tissot BP, Welte DH (1984) Petroleum formation and occurrence–second revised and enlarged edition

57. Versteegh GJ, Blokker P (2004) Resistant macromolecules of extant and fossil microalgae. Phycol Res 52(4):325–339

58. Wojdyr M (2010) Fityk: a general-purpose peak fitting program. J Appl Crystallogr 43(5–1):1126–1128

59. Yang J, Hatcherian J, Hackley PC, Pomerantz AE (2017) Nanoscale geochemical and geomechanical characterization of organic matter in shale. Nat Commun 8(1):2179

60. Yang A (2016) Experimentally quantifying the impact of thermal maturity on aromaticity and density in Kerogen (M.Sc. thesis)

61. Zeszotarski JC, Chromik RR, Vinci RP, Messmer MC, Michels R, Larsen JW (2004) Imaging and mechanical property measurements of kerogen via nanoindentation. Geochim Cosmochim Acta 68(20):4113–4119

Chapter 4
Understanding Organic Matter Heterogeneity and Maturation Rate by Raman Spectroscopy

Abstract Solid organic matter (OM) in sedimentary rocks produces petroleum and solid bitumen when it undergoes thermal maturation. The solid OM is a 'geomacromolecule', usually representing a mixture of various organisms with distinct biogenic origins, and can have high heterogeneity in composition. Programmed pyrolysis is a common method to reveal bulk geochemical characteristics of the dominant OM, while detailed organic petrography is required to reveal information about the biogenic origin of contributing macerals. Despite the advantages of programmed pyrolysis, it cannot provide information about the heterogeneity of chemical compositions present in the individual OM types. Therefore, other analytical techniques such as Raman spectroscopy are necessary. In this study, we compared geochemical characteristics and Raman spectra of two sets of naturally and artificially matured Bakken source rock samples. A continuous Raman spectral map on solid bitumen particles was created from the artificially matured hydrous pyrolysis residues, in particular, to show the systematic chemical modifications in microscale. Spectroscopic data was plotted for both sets against thermal maturity to compare maturation rate/path for these two separate groups. The outcome showed that artificial maturation through hydrous pyrolysis does not follow the same trend as naturally-matured samples although having similar solid bitumen reflectance values (%SBRo). Furthermore, Raman spectroscopy of solid bitumen from artificially matured samples indicated the heterogeneity of OM decreases as maturity increases. This may represent an alteration in chemical structure towards more uniform compounds at higher maturity. This study may emphasize the necessity of using analytical methods such as Raman spectroscopy along with conventional geochemical methods to better reveal the underlying chemical structure of OM. Finally, observation by Raman spectroscopy of chemical alteration of OM during artificial maturation may assist in the proposal of improved pyrolysis protocols to better resemble natural geologic processes.

Keywords Organic matter · Heterogeneity · Raman spectroscopy · Maturation rate · Hydrous pyrolysis

4.1 Introduction

Shale is the most abundant fine-grained sedimentary rock and is formed from compaction of silt and clay-sized minerals which also may contain a significant amount of solid organic matter (OM). Kerogen (the insoluble portion of OM) is a macromolecule and a mixture of OM types with different origins (Types I, II, III and IV) [17]. When kerogen experiences maturation, weight concentration and the composition of the individual macerals evolve [79]. During this process, bigger molecules will break down and the outcome will be hydrocarbons and bitumen in addition to water, CO_2 etc. [66]. The remainder of the organic matter acts as the storage reservoir by developing nanoscale pores which hold generated hydrocarbons [8, 11]. Furthermore, the process of thermal maturation will cause or enhance the development of local heterogeneities within the remaining organic matter [35]. This is due to differences in the rate of maturation for the existing kerogen based on differences in type and origin [79]. Additionally, heterogeneity may also be due to local variations in thermal alteration that is imposed on organic matter by catalysis from nearby mineral grains [59]. Understanding these heterogeneous patterns will enable better characterization of geochemical, petrophysical and geomechanical properties of organic matter from a molecular point of view [11]. In addition, these studies will enhance our understanding about mechanisms related to generation and migration of hydrocarbons, especially in self-sourced unconventional reservoirs.

Despite the importance of assessing shale OM heterogeneity, common bulk-rock geochemical measurement methods that have been utilized for decades such as programmed pyrolysis and LECO total organic carbon (TOC) are incapable of providing information specific to individual OM types. SEM (scanning electron microscope) analysis combined with energy-dispersive X-ray spectroscopy (SEM–EDS) can detect local changes in the abundance of higher atomic weight elements. However, it cannot recognize variations in molecular chemical components which are the driving force behind OM heterogeneity [61, 69]. Organic petrography allows identification of various types of organic components but cannot detect molecular compositions [20]. Moreover, geochemistry of organic matter which is controlled by molecular compounds is usually investigated by pyrolysis which is artificially maturing the samples. Yet, to obtain meaningful results, many parameters (pyrogram) that are involved in the experiments should be set accurately to resemble a natural maturation path [7]. This process can be done in the absence or presence of water, anhydrous pyrolysis (AHP) and hydrous pyrolysis (HP), respectively.

Lewan et al. [39] discussed and performed several experiments on the role of water in petroleum formation. He showed that AHP will result in higher organic maturation rate compared to HP at the same temperature conditions. Michels et al. [52] discussed the effects of effluents and water pressure on oil generation during confined pyrolysis. Lewan and Ruble [38] showed that there is no correlation between OM kinetic parameters derived from open-system pyrolysis and HP. Pan et al. [59, 60] illustrated the impact of water on the organic carbon ratio and mineral acidity during the conversion process of OM to hydrocarbons. They concluded that organic matter maturation rate slows down when a large amount of water is present. Further, Lewan and Roy [37] explained significant differences in the types of products when water is involved at higher thermal maturity, noting that water promotes thermal cracking of bigger molecules over cross-linking. Other researchers have investigated the role of clay minerals in maturation, crude oil formation, migration and accumulation [8, 11, 22, 26, 78, 84]. Such studies illustrated that the composition of petroleum is modified by interaction of clay minerals, and also the presence of water affects the role of minerals in acid-catalyzed cracking of kerogen. Despite advances in our understanding of these interactions, mimicking the natural condition of kerogen conversion to hydrocarbon remains poorly understood.

In a molecule, atoms are connected by chemical bonds, and thus have periodic motions. These motions relative to each other are superpositions of normal mode vibrations with the same phase and normal frequency [68]. The most effective methods to observe vibrational spectra as a signature representing a specific chemical compound are infrared and Raman spectrometry.

Kelemen and Fang [27] showed the application of Raman spectroscopy for studying thermal maturity of OM, particularly in Silurian and older rocks where vitrinite is absent. Other researchers also have used Raman spectroscopy to reveal structural changes in organic matter during maturation [16, 18, 24, 43, 63]. Khatibi et al. [32, 28, 30, 31] showed the potential application of Raman spectroscopy in correlating Raman signals to geochemical and mechanical properties of organic matter. They explained that changes in Young's modulus are representative of molecular alterations happening through thermal maturation. The application of Raman spectroscopy in qualitatively predicting geochemical properties of organic matter in terms of Rock–Eval parameters has also been studied [32, 28]. Although Raman spectroscopy suffers from strong fluorescence background noise for immature shale samples, Khatibi et al. [31] showed that solvent extraction will reduce the fluorescence background if combined with specific signal processing techniques.

Khatibi et al. [29] applied Raman spectroscopy to investigate OM heterogeneity within the organic matter, solid bitumen, in naturally and artificially matured source rock samples from the Bakken Formation. They conducted OM reflectance measurement, programmed pyrolysis and Raman spectroscopy to compare the maturation pathways of the naturally and artificially matured samples. They collected six samples from the source members (upper and lower) of the Bakken Formation for their study. These two organic-rich shale members were classified as the same lithofacies type by some authors (e.g., [3, 82]) and are considered to have been deposited under relatively deep marine (> 200 m depth) anoxic conditions [71]. The samples were at different stages of thermal maturity varying from 0.38 to 0.98%SBR$_o$ (Solid Bitumen reflectance) along with one immature sample (%SBR$_o$ of 0.32) that was artificially matured by hydrous pyrolysis (HP) to maximum maturity of 1.29%SBR$_o$. The HP process followed by Khatibi et al. [29] involved isothermal temperature treatment (300–360 °C) of the homogenized immature sample based on the method of Lewan [40, 36]. Crushed rock (homogenized) samples (2–4 g) were loaded into reactors and covered with de-ionized water in a way that the amount of water should cover the sample and be in contact with it constantly at the maximum temperature for each step of the experiment. The amount of rock and water added to each experiment were calculated from rock densities, reactor internal volumes, and steam tables as described by Lewan [40]. Table 4.1 shows the results of geochemical analyses performed on the artificial and naturally matured samples. For organic petrographic analysis, the samples were prepared according to ASTM D2797 wherein they mounted the rock particles in a thermoset plastic briquette, then ground and polished with successively finder abrasives until a 0.05 μm finishing stage. For solid bitumen reflectance analyses they followed ASTM D7708. In this technique, incident white light 546 ± 10 nm is reflected from solid bitumen positioned under the microscope crosshairs at 500× magnification, measured at a detector and compared to measured light reflected from a calibration standard. They collected 20 measurements of solid bitumen reflectance for each sample, with only 1 measurement per individual rock fragment. Since there was no reliable identification of primary vitrinite in the samples, their study [29] reported mean random solid bitumen reflectance (%SBRo) reported in place of vitrinite reflectance. All the analysis in their study were conducted on solid bitumen for better comparison purposes and consistency. Solid bitumen and vitrinite in shale are generally characterized as structureless gray substances in reflected white light (Fig. 4.1).

4.2 Raman Spectroscopy

To generate the Raman spectrum, Khatibi et al. [29] in their study illuminated the sample surface with monochromatic laser light. Raman mapping mode was acquired on the region of interest (ROI) where organic matter (solid bitumen) was detected. As the probe scanned the area, 1000 points were collected (2 spectra per 1 μm^2) (Fig. 4.2. The power of laser that they used for the acquisition of Raman data was 1 mW (1% of total laser power, with 1 s acquisition time, 3 s of accumulation which led to 3 mW s of the energy dumped onto the surface of the sample. In terms of spectral precision, the grating (1200 grating/mm) was not moved from one spot to another; therefore, there was not any change in the spectral calibration. Moreover, the Raman spectrometer was calibrated before each data acquisition session using the 520 cm^{-1} Raman band of a silicon reference sample. The Si Raman band was used as an adjustment to an overall calibration of the spectrometer made with atomic emission lamps over the entire working spectral range of the spectrometer.

One of the most important challenges regarding Raman spectroscopy is the fluorescence background which masks the Raman signal [57, 83]. This issue is more critical while evaluating lower maturity samples due to the presence of high fluorescence intensity in OM, which is reduced at higher maturity [47, 65]. An appropriate processing technique can notably reduce the fluorescence background levels. For this purpose, Khatibi et al. [29] in their study fitted a polynomial baseline curve to the whole spectrum (from 50 cm^{-1} to 2750 cm^{-1}) (Fig. 4.3a), which was the long

Table 4.1 Geochemical properties of two sets of Bakken samples: artificially matured by HP and naturally matured. Note the systematic changes in T_{max}, TOC, S2, HI, and PI with increasing maturity in both naturally and artificially matured samples

	Sample No.	Depth (ft)	TOC (wt%)	S1 (mg HC/g)	S2 (mg HC/g)	Tmax (°)	HI (S2 * 100/ TOC)	OI (S3 * 100/ TOC)	P1 S1/ (S1 + S2)	SBR$_O$ (%)	HP time (h)	HP temp. (°C)
Artificially matured sample by HP	HP1	7652	14.59	8.28	80.83	428	554	8	0.09	0.32	72	Original
	HP2		13.3	4.50	72.02	429	542	10	0.06	0.35	72	280
	HP3		15.15	8.00	87.1	434	575	7	0.08	0.38	72	300
	HP4		11.55	7.85	72.65	436	629	12	0.1	0.49	72	310
	HPS		10.66	6.78	53.47	431	502	9	0.11	0.54	72	320
	HP6		7.35	4.18	18.47	442	250	11	0.18	0.95	72	340
	HP7		8.12	3.19	11.14	451	137	12	0.22	1.19	72	350
	HP8		7.95	3.97	8.07	462	102	6	0.33	1.29	72	360
In situ matured samples	Well 1	5438	24.73	7.97	128.71	419	520	8	0.06	0.38	–	–
	Well 2	8326	16.27	8.27	90.69	428	557	2	0.08	0.54	–	–
	Well 3	9886	15.76	9.27	83.7	432	531	1	0.1	0.59	–	–
	Well 4	10,555	13.26	0.31	33.03	449	260	1	0.01	0.86	–	–
	Well S	10,725.5	9.04	6.13	13.94	450	154	1	0.31	0.94	–	–
	Well 6	11,199	16.36	0.71	28.05	452	171	1	0.02	0.92	–	–

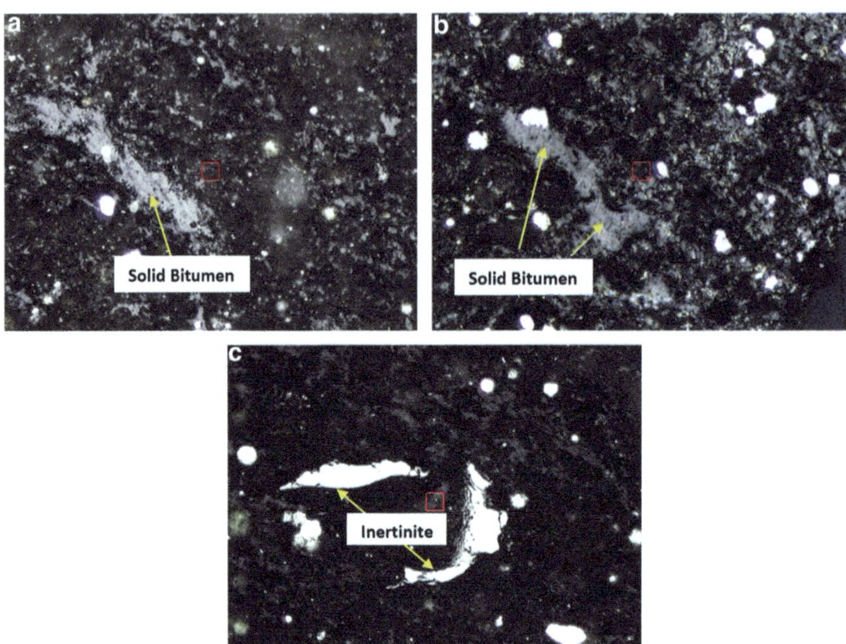

Fig. 4.1 **a** and **b** Solid bitumen in samples of Well 1 and Well 5, respectively, **c** occurrence of dispersed organic matter in the form of inertinite in Well 4. The red square in the middle of each image is a scale of 5 μm of each side

wavelength function of the fluorescence, and then was subtracted from the spectrum (Fig. 4.3b). Lower order polynomial functions are less flexible to adapt with the data and may not be able to include highly fluctuating backgrounds. Use of this mathematical approach provides a flat zero baseline spectrum to pick Raman band positions in a more straightforward manner.

To improve signal to noise ratio (S/N) during Raman acquisition a binning procedure can be used [29]. The binning procedure involves co-adding of a number of adjacent spectra to improve the signal to noise ratio. Khatibi et al. [29] to increase the S/N ratio as shown in Fig. 4.4. However, they noted that while using the binning factor one should be cautious to avoid losing spectral features because of large binning.

To characterize Raman spectral parameters such as peak position, intensity and area, a curve fitting step is performed. Researchers often fit 5 peaks for OM evaluation in shale plays [9, 46, 65–67], known as D, G, D2, D3 and D4 as shown in Fig. 4.5. Khatibi et al. [29] in their study also fitted 5 peaks, but only considered the major bands (D and G) in the results (Table 4.2) as the molecular significance and appearance of the minor bands (D2, D3 and D4) are not well understood [6].

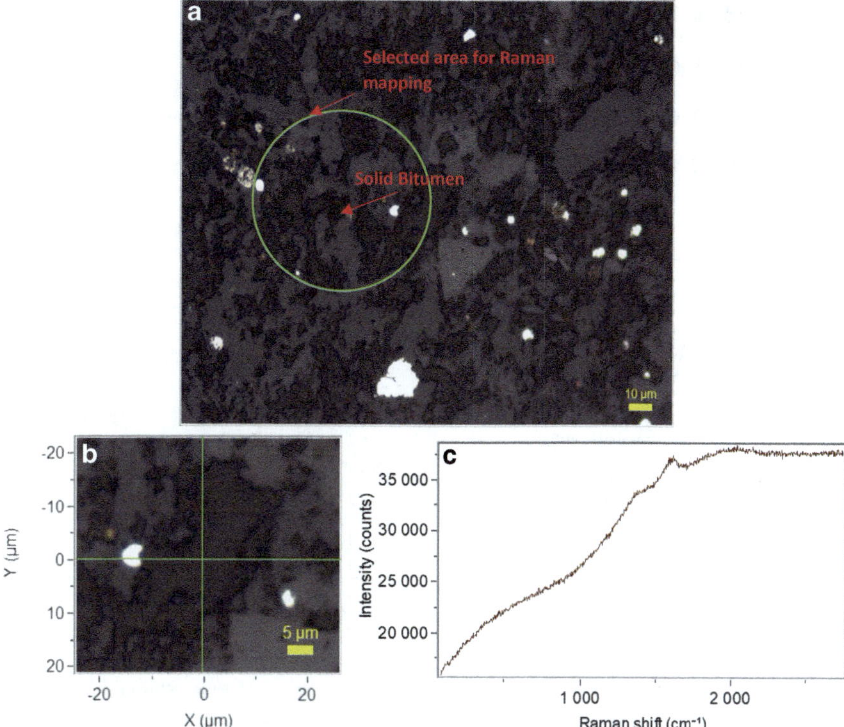

Fig. 4.2 **a** Area selected for Raman spectroscopy, a complete spectrum is acquired at every pixel of the area of interest (1000 spectra are acquired which can all be processed with the same processing steps), **b** a point selected on solid bitumen, **c** and its corresponding Raman spectrum

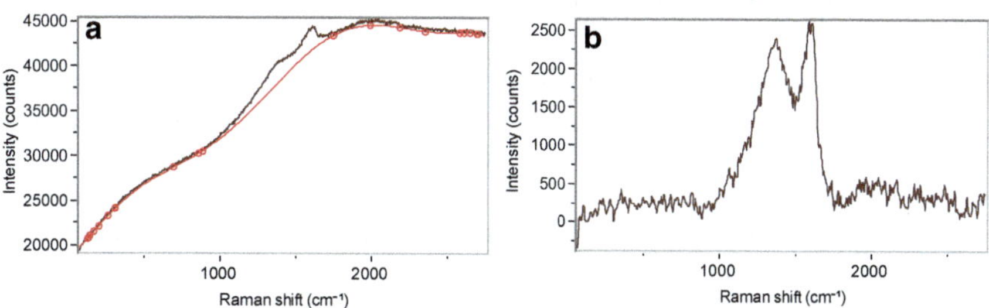

Fig. 4.3 Raman signals before (**a**) and after (**b**) baseline correction for Well 2 as a naturally matured sample

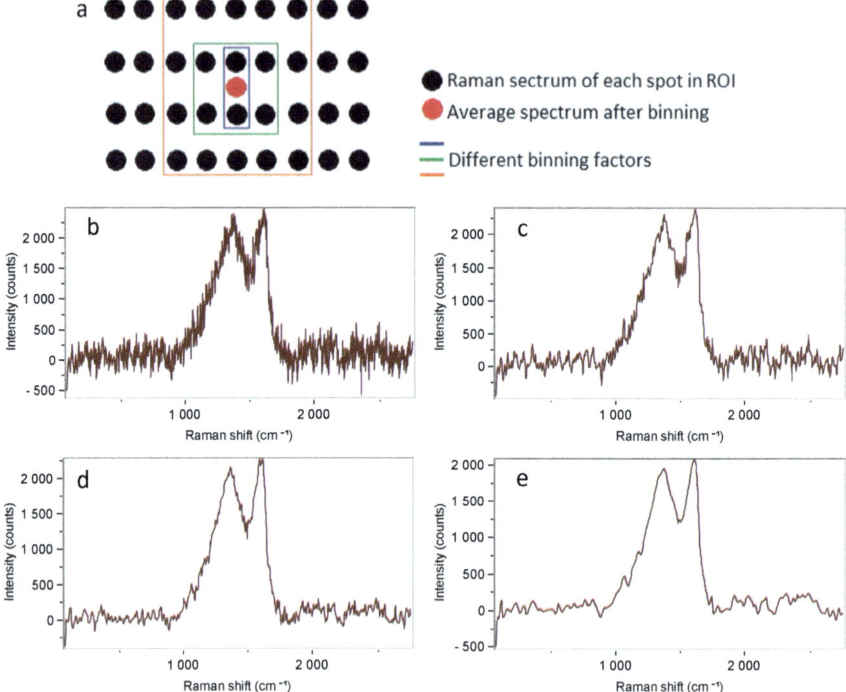

Fig. 4.4   **a** Schematic of binning; **b** single spectrum; **c** binning factor of 4; **d** binning factor of 8; **e** binning factor of 20. Note the increase in signal to noise ratio and correspondingly losing spectral features by increasing binning factor

4.3   Evolution of OM Thermal Maturity Pathways

Several studies have reported that thermal maturity causes a logical variation in the positions, separations, and other important parameters of Raman spectral bands for organic matter [16, 27, 28, 32, 65, 70]. In Raman spectra, the major bands of D and G represent the aromaticity level of organic matter and the level of disorder in molecular structure of organic atoms. Therefore, Raman spectroscopy can detect different levels of thermal advance based on the response from these molecular signals. A well-known increase in the G-D band separation with maturity (up to dry gas window) is due to the shift in the D band position towards lower wavenumbers and the G band towards higher wavenumbers [27]. These shifts are attributed to the increase of larger aromatic clusters and better ordered-structure kerogen in terms of the existing organic compounds [66]. Khatibi et al. [29] in their study demonstrated band separation through cross-plots of SBRo versus band separation and T_{max} versus band separation for both naturally and HP samples (Fig. 4.6). They observed that functions fitted to the two separate datasets (naturally and artificially matured samples) show visibly distinct slopes. It is known that different organic matters mature at various rates

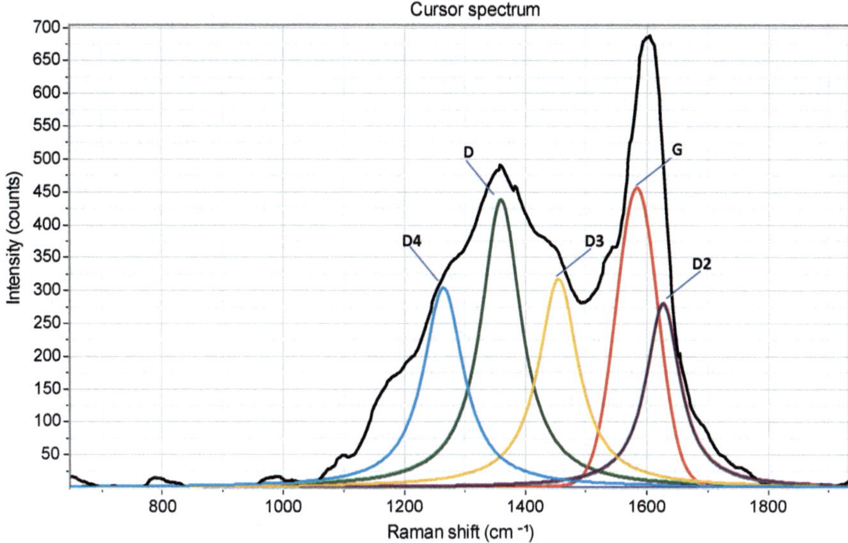

Fig. 4.5 Schematic of band peaking procedure using 5 peaks (D, G, D2, D3 and D4) and the experimental Raman spectrum (black). By adding more peaks, the reconstructed spectrum is more similar to experimental results and the χ^2 value is reduced

Table 4.2 Major D and G bands shifts for both artificially and naturally matured samples in this study derived from the average spectrum across the ROI

	Sample No.	D band	G band	G-D band
Artificially matured sample by HP	HP1	1362	1600	238
	HP2	1363	1598	235
	HP3	1362	1601.5	239.5
	HP4	1358	1603	245
	HP5	1359	1602	243
	HP7	1363	1607	244
	HP8	1357	1604	247
Naturally matured samples	Well 1	1367	1585	218
	Well 2	1367	1587	220
	Well 3	1360	1591	231
	Well 4	1354	1591	237
	Well 5	1359	1594	235
	Well 6	1357	1601	244

due to presence of specific functional groups [79] with different kinetic parameters. To remove this effect, they exclusively used solid bitumen in their study. Lewan and Ruble [38] concluded kinetic parameters derived from HP would determine geologically significant differences between source rocks bearing different kerogen types, conversely to open-system pyrolysis. This difference in maturation trend is also observed by Khatibi et al. [29] for changes in bulk measurements such as TOC, S2 and HI, as shown in Fig. 4.7. Relationships between measured %VRo and programmed pyrolysis values have been introduced in previous studies [19, 34, 58]. For example, Epistalié et al. [15], Dembicki [12] and Abarghani et al. [1] investigated different geochemistry datasets from the Bakken and presented a 'normal' kerogen maturation pathway for T_{max} as a function of %VRo as shown in Fig. 4.8. Figure 4.8 also depicts artificially matured samples of Khatibi et al. [29] do not follow the trend of natural ones in T_{max}-VRo space which was also reflected in the Raman spectra (Fig. 4.6). This difference was mainly inferred to the fact that HP residues are still reactive. Therefore, although the SBRo of HP samples were the same as the natural sequence, their compositions were not.

The discrepancy in the rate and path of maturation for naturally and artificially matured samples can be referred to as a separate reaction medium between HP and natural geologic processes of maturation. The importance of reaction medium (defined as the effect of water, minerals, heating rates, and pressure on organic maturation in HP) has been discussed extensively [4, 13, 42, 49, 51, 53, 54, 59, 60]. Water as the exogenous source of hydrogen seems to have an important role in oil generation and simulation of the natural reaction medium [36], whereas the dominant reaction pathway in the absence of water is formation of pyrobitumen due to C–C bond cross linking. Additionally, in the presence of water, the primary reaction pathway is formation of saturate-enriched oil due to thermal cracking of C–C bonds [55]. Mineral acidity also increases maturation rate as well as affecting the byproducts from maturation [59], while impermeability of fine-grained sediments prevents forward reaction progress [21, 50] (Li et al. 2004). Several researchers have investigated the effects of reaction medium in formation rates of hydrocarbon [10, 73]. Based on the extent of kerogen conversion, the amount of trapped bitumen in the pyrolyzed and naturally matured samples might be also different. Monthioux et al. [55] showed natural maturation is best simulated when pyrolysis is performed under confined conditions in which free volume or diluting inert gas is not present. Kinetic constants of geochemical reactions derived from HP has shown that reaction mechanisms in nature and in the laboratory are distinctly different [48, 64]. Therefore, the presence of water, minerals, heating rates, and pressure affects the maturation paths of organic matter in addition to the variability of the macerals present. Kinetics of hydrocarbon generation and maturation all are dependent upon molecular structure of kerogen [5, 33, 72]. This dependency can also be extended to nanomechanical and nanoporosity evolution of organic matter [8, 44, 45]. In this regard, bulk chemical measurements suffer notably from representing molecular structure of kerogen, a property which is not known fully. For instance, although Rock–Eval data and

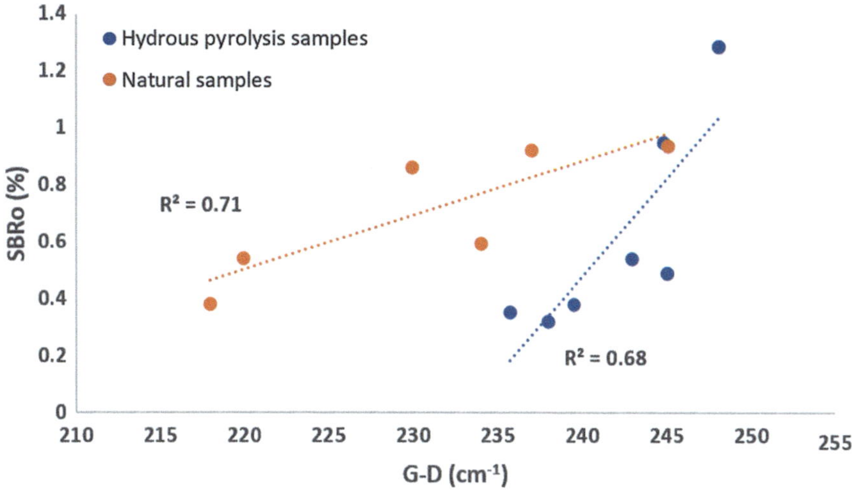

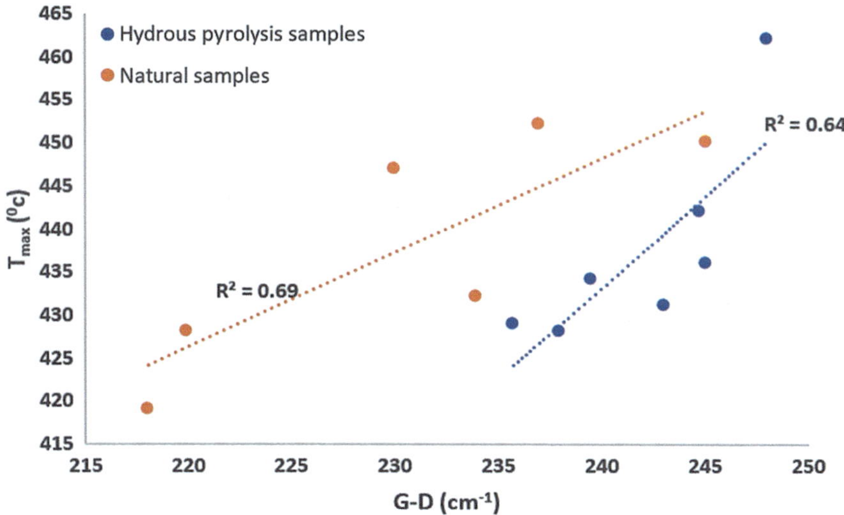

Fig. 4.6 SBRo and T_{max} (as maturity indicators) versus band separation. Note the consistent increase of band separation with increasing thermal maturity. Trendlines for HP and naturally matured samples along with their corresponding correlation coefficients are shown on each plot

(Pseudo) Van Krevelen diagram as seen in Fig. 4.9, exhibited very similar organic matter type/origin, detailed molecular analyses of the samples are proved otherwise.

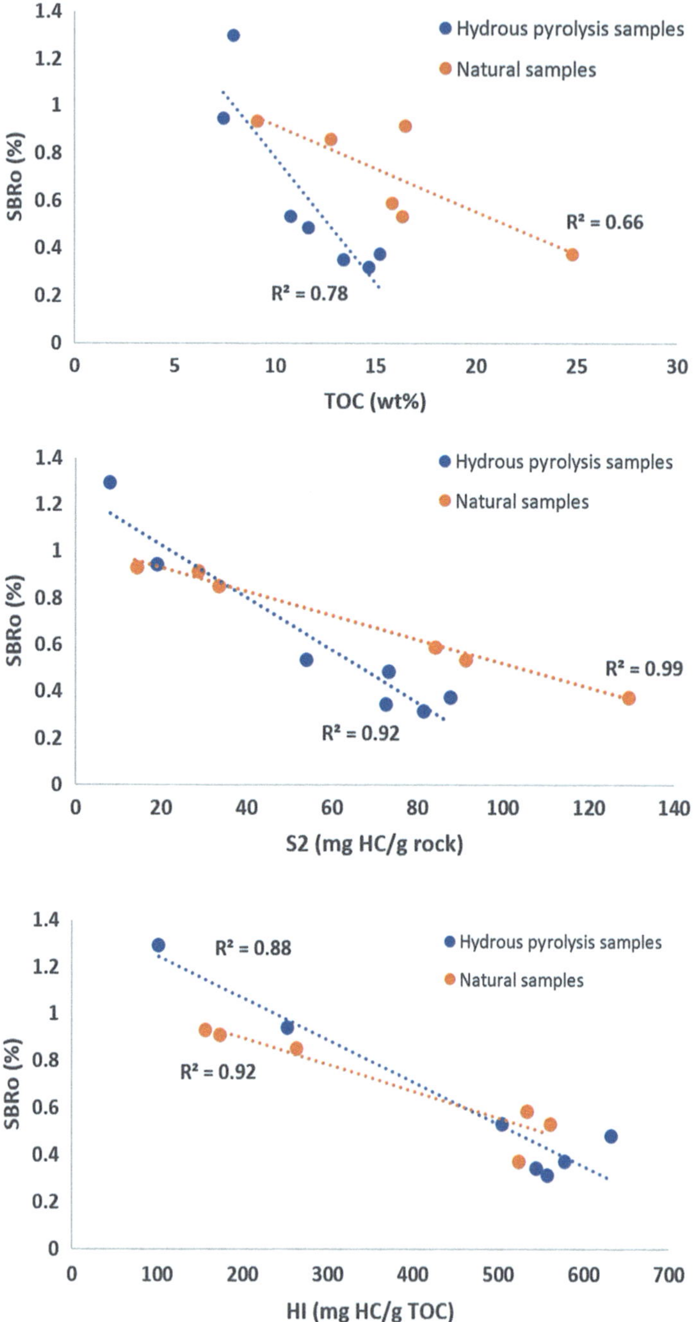

Fig. 4.7 (Top) %SBRo versus TOC; (middle) %SBRo versus S2; (bottom) %SBRo versus HI. For each parameter plotted on the abscissa, the trend versus %SBRo is different for artificially and naturally matured samples

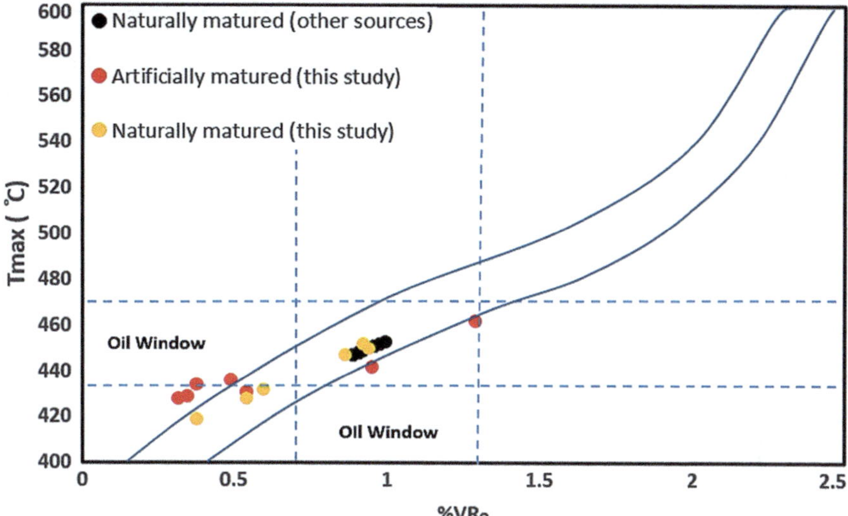

Fig. 4.8 Observed relationship between VRo% and Tmax for natural samples (modified after Epistalié et al. [15]; [1, 12]). It should be noted that measurements were based on %SBRo and were converted to vitrinite reflectance %VRo using the [23] equation. Hydrous pyrolysis results (red circles) do not exactly follow the natural trend limits with dark blue curves. Light orange circles are naturally matured samples in this study, and black circles are naturally matured Bakken data extracted from Abarghani et al. [1]

4.4 OM Heterogeneity

To study OM heterogeneity using Raman spectroscopy, Khatibi et al. [29] selected regions on the surface of solid bitumen and their corresponding spectra were analyzed, as presented in Figs. 4.10, 4.11 and 4.12. To avoid the inconsistency of different biogenic sources and natural variability, the following results are based on HP samples and Raman spectra were obtained only from solid bitumen particles. To minimize the effects of technical detail (excitation laser energy, processing steps, sample preparation etc.), the same procedures were used for all samples.

It has been shown that carbonaceous matters are sensitive to the polishing process, and the effect is significant [47]. The excitation wavelength might also affect the Raman results, so to avoid such bias, the same wavelength for each and all sample surfaces was used to acquire Raman spectra. Moreover, short laser excitation wavelengths (488 nm and less) are usually used for Raman acquisition of organic matter to invoke far less fluorescence background noise [47], which was also the case in this study. Raman spectral mapping mode showed that different locations within the same solid bitumen particle did not necessarily present the same Raman spectral characteristics. This variation in the spectra is represented herein by G-D band separation (Figs. 4.10, 4.11 and 4.12). If it is assumed that polishing and laser wavelength had negligible effects on collected data, it can be concluded that the variability in

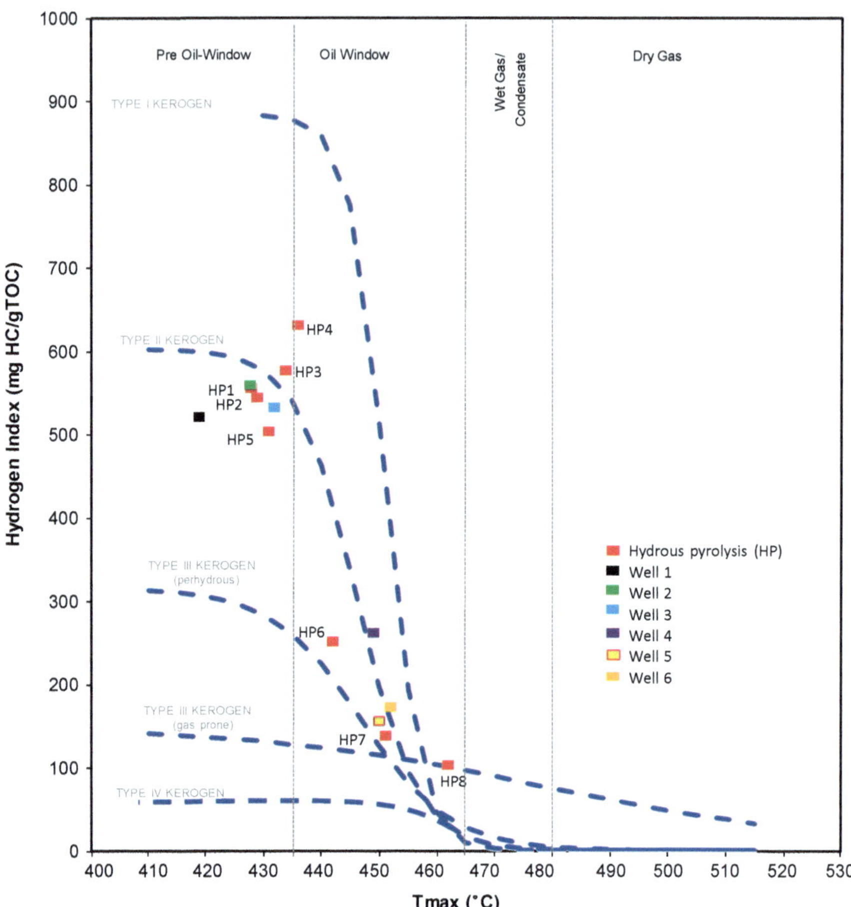

Fig. 4.9 Hydrogen index versus T_{max} for kerogen typing. As seen all samples including naturally and artificially matured are approximately in type II kerogen (modified after Tyson [76])

Raman spectral results might be due to the heterogeneity of organic matter in terms of chemical composition. Results are also in agreement with recent research that was performed by Jubb et al. [25] who detected high variability in the Raman response across a less than 5 μm spatial distances of single OM in shale samples.

Moreover, as thermal maturity advances, observed variation in G and D band parameters reduce for different locations within the same solid bitumen particle. This observation suggests that heterogeneity of organic matter decreases with increasing thermal maturity (note the standard deviation values for different samples in Figs. 4.10, 4.11 and 4.12). Similar results are also reported by Jubb et al. [25] for Niobrara shale sample that OM chemical heterogeneity is lost for samples with higher thermal maturity. This can be explained by increase in aromaticity, and heteroatoms expulsion, therefore, no matter its origin or type, OM tends to become graphite-like

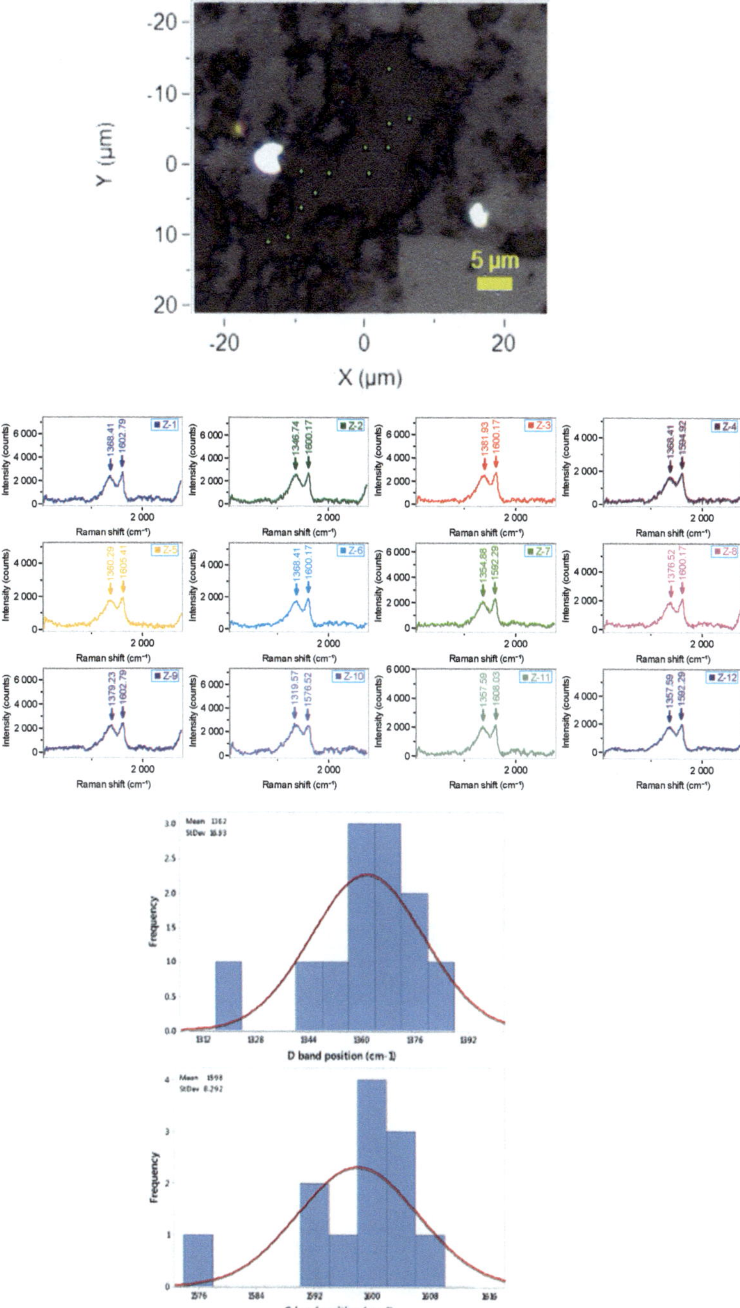

Fig. 4.10 Sample HP1 with 0.32%SBRo: (top) Showing spots selected for averaging Raman signals (each green circle has about 3 3.6 μm^2 area); (middle) Raman spectra of corresponding spots; (bottom) Histogram of D and G bands position

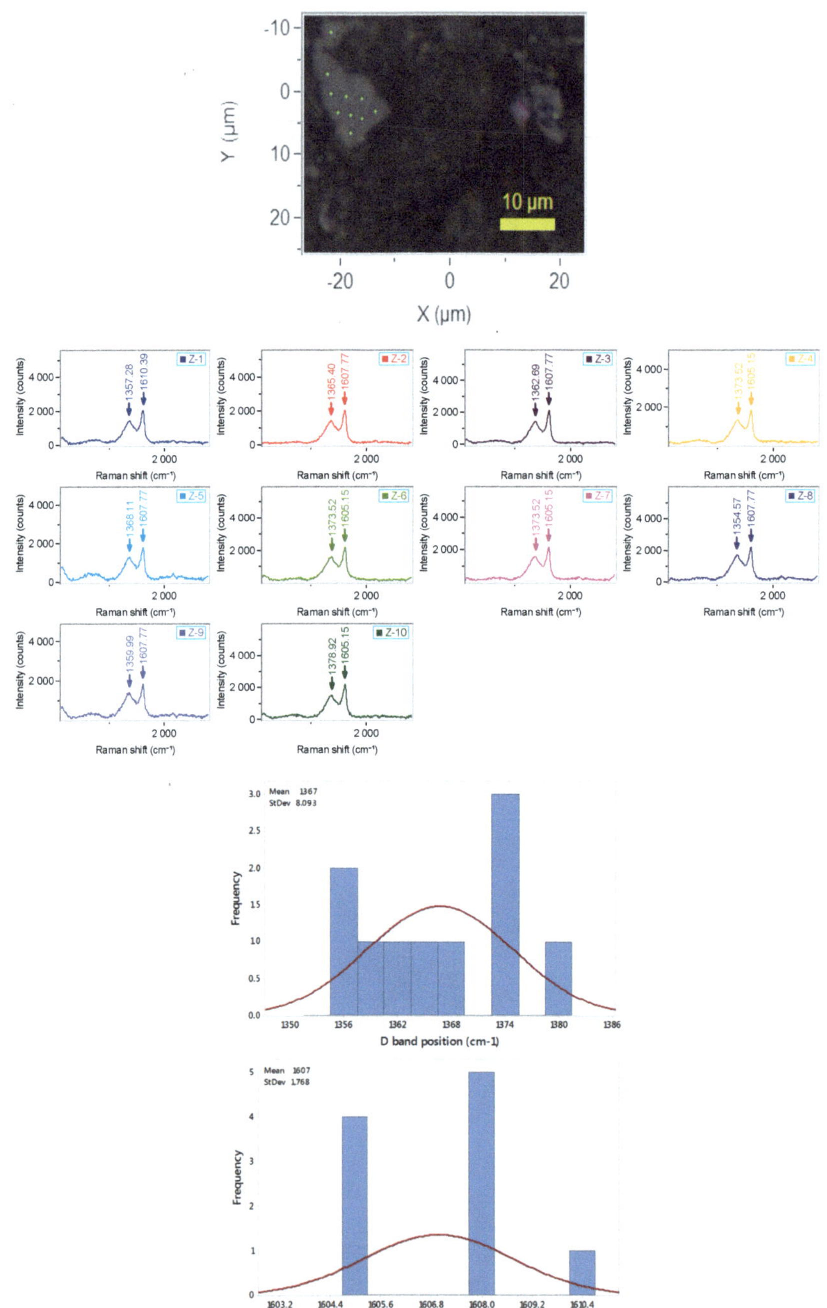

Fig. 4.11 Sample HP6 with 0.95%SBRo: (top) Showing spots selected for averaging Raman signals; (middle) Raman spectra of corresponding spots; (bottom) Histogram of D and G bands position

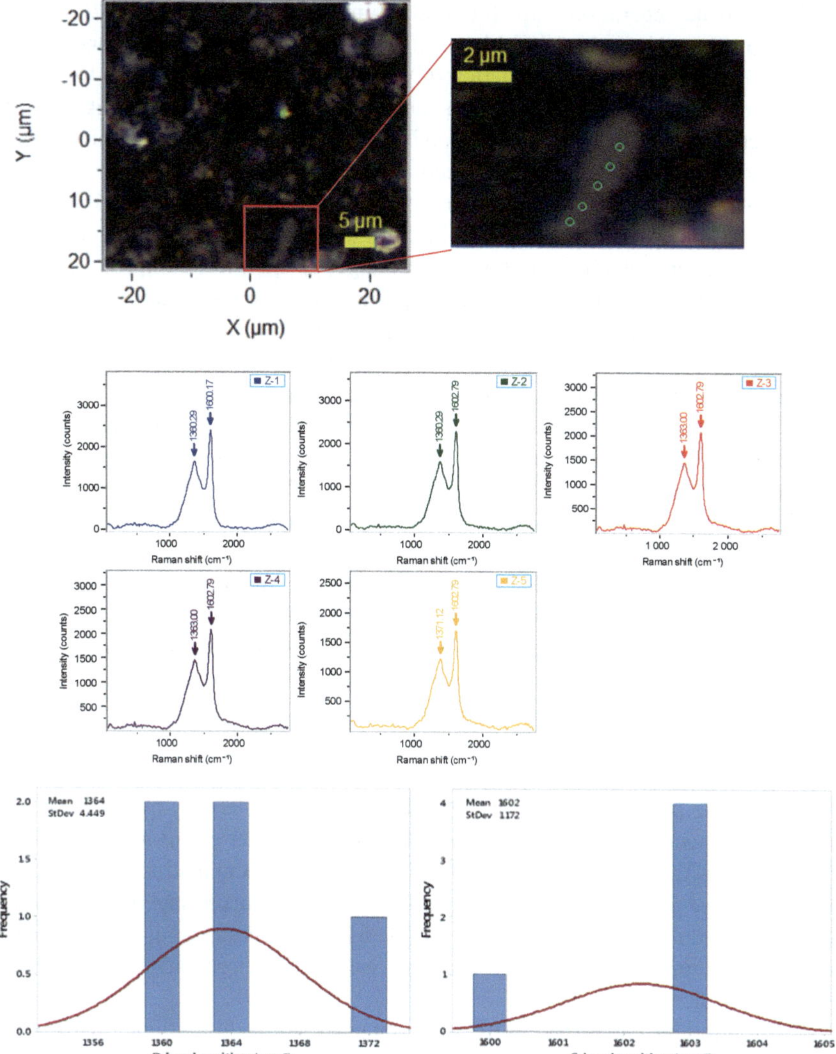

Fig. 4.12 Sample HP8 with 1.29%SBRo: (top) Showing spots selected for averaging Raman signals; (middle) Raman spectra of corresponding spots; (bottom) Histogram of D and G bands position

at higher maturity [16, 47, 56, 62, 63, 66, 74–77, 83]. This is also in accordance with the original Van Krevelen diagram in which all kerogen types converge to the origin as their composition and structure become similar when thermal maturity is increased [33].

Raman band separation maps for solid bitumen particles in Figs. 4.10, 4.11 and 4.12 are displayed in Fig. 4.13. Table 4.3 also shows the corresponding standard deviation (SD) and coefficient of variation (CV) of spectra for each sample. Results are showing decreasing of band separation variability with increasing maturity. Interpolation method for surface data generation is performed using the minimum curvature method which uses an operator that keeps the surface data smooth with a minimum amount of bending, while attempting to honor the data as closely as possible [2, 80]. It can be seen, across the surveyed solid bitumen particles, band separation (G-D) exhibits variations which might represent changes in chemical composition, matrix effects (specifically near the edge of the grain) and/or surface quality. In order to demonstrate these variations in Raman signals across each surveyed particle more clearly, CLS (classical least square) fitting was used as a multivariate decomposition technique. This procure was used to calculate the contribution of the reference spectra in the same size area to create a profile based on the similarity of each spectrum to the reference spectrum. The reference spectrum was selected in the middle of each sample to be far from edges, Fig. 4.14. From the figures it can be concluded: by increasing maturity the magnitude of fluctuations in band separation is decreasing, and distribution of band separation will become more uniform throughout the solid bitumen particle which might denote advancement towards compositional homogeneity. Similar results were observed by Lünsdorf [47] and Jubb et al. [25].

This representation of heterogeneity within a solid bitumen phase may explain other heterogeneous behaviors such as mechanical properties of OM. It has been shown in the literature, as thermal maturity increases, mechanical properties (e.g., elastic modulus) of organic matter alter (Eliyahu et al. 2014; [41]). It can be described as in immature source rocks the OM appears to surround other minerals, becoming a load-bearing part of the rock framework [44, 81]. Conversely, by increasing the maturity, kerogen becomes more isolated between the other grains and results in an increase in its Young's modulus [14, 81]. In terms of molecular structural evolution, organic matter transforms gradually from chaotic and mixed-layered to a better-ordered aromatic molecular structure [6, 31, 63]. This reasoning is also consistent with the previous paragraph regarding decreasing of heterogeneity of molecular structure by increasing maturity.

The continuous chemistry information obtained from solid bitumen particles by Raman spectroscopy may have the potential to enhance conventional petrographic or bulk geochemistry analytical approaches. For example, Raman spectroscopy has the benefit of minimal sample preparation (It can be done on sample chips as well) in a non-destructive manner and fast acquisition by pinpointing the organic matter and illuminating by proper laser wavelength. Additionally, the cross-plot of SBRo with Raman data, can enable us to adjust the condition of pyrolysis to make it resemble natural maturation pathways more accurately at the small scale and confirm it with bulk measurements in programmed pyrolysis. To do so, several HP and AHP setups can be performed and the cross-plot of Raman bands versus maturity should be compared to naturally matured samples. Moreover, detecting heterogeneity and providing maturity maps can also be a tool for further understanding of kerogen characteristics. The latest requires extensive sample collection (sample number/ft) and

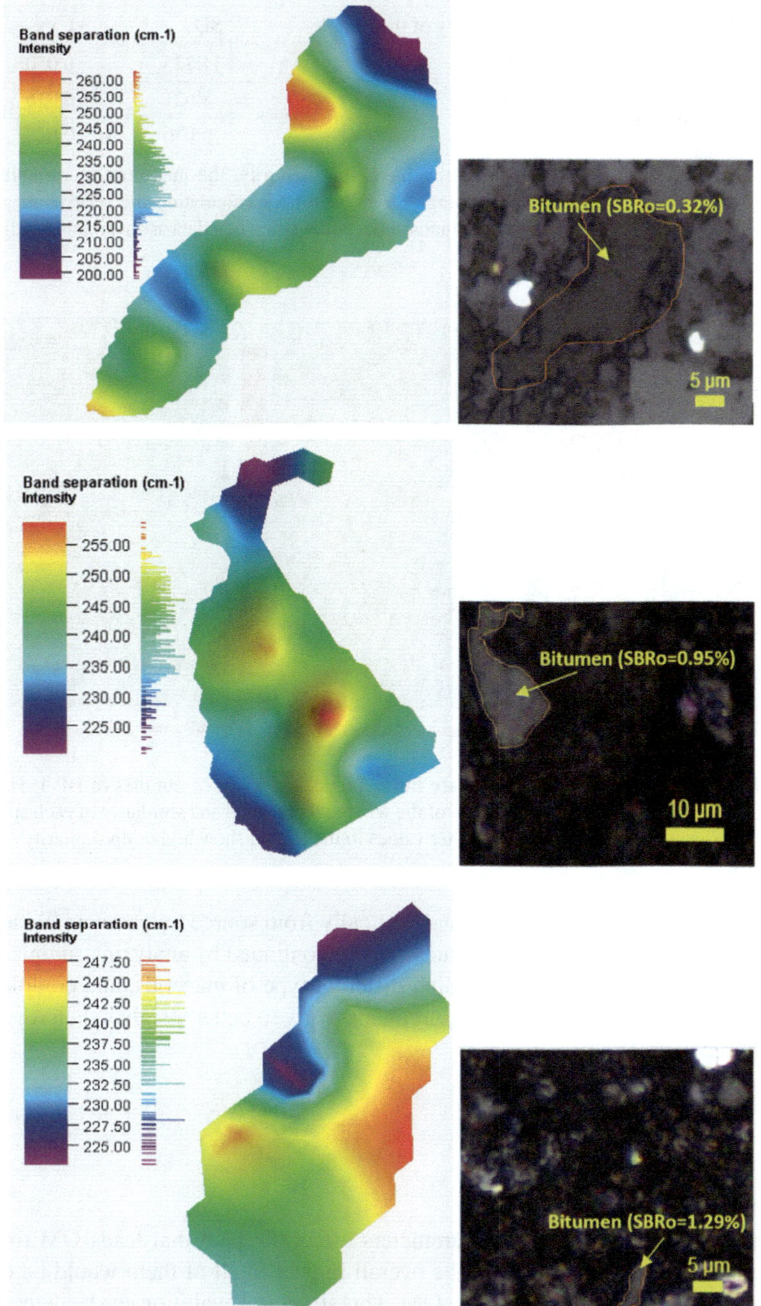

Fig. 4.13 Raman map of a bitumen corresponding to Figs. 4.10, 4.11 and 4.12 (HP 1, HP6 and HP8, respectively). Note the changes of band separation indicating changes of chemical composition of the solid bitumen. As seen, variation of band separation is reducing by increasing maturity. Please note bar graphs next to color scales shows the frequency of each band separation

Table 4.3 Standard deviation (SD) and Coefficient of variation (CV) of band separations for HP 1, HP6 and HP8 for the entire area of interest

Maturity of the sample	SD	CV
0.32	11.725	0.050
0.95	8.621	0.036
1.29	6.036	0.026

As see, by increasing maturity, the measures of variability are decreasing. CV is useful for calculating the relative magnitude of the standard deviation when few data are available for different datasets

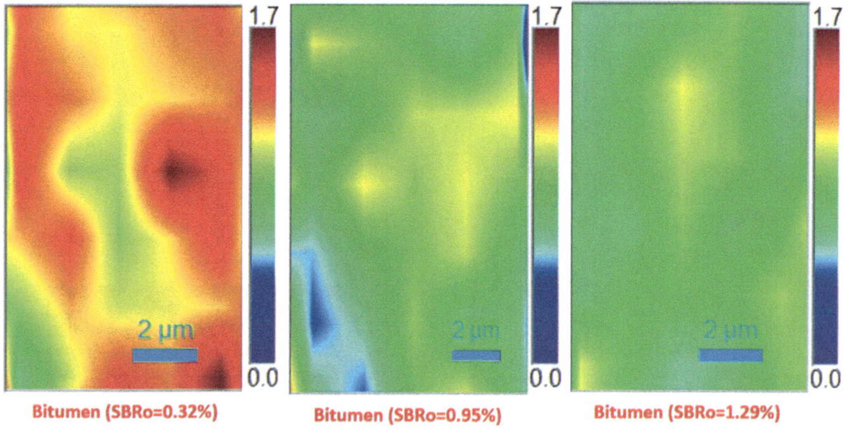

Fig. 4.14 Results of classical least square fitting procedure for three samples of HP 1, HP6 and HP8. A reference spectrum in the middle of the window is selected and similarity of each spectrum to the reference spectrum is shown. Higher values in the legend show higher dissimilarity

spectroscopy study both laterally and vertically from source to reservoir. To accomplish more extensive results, this study will be continued by analyzing samples with a wider range of maturities, including different type of macerals, and complement our Raman data with Nano-IR measurements to give a better insight to heterogeneity of organic matters.

4.5 Conclusion

It was understood that various parameters affect the path that leads OM towards higher maturity levels, however, the overall impact of all of them would be on the chemical structure of the remaining OM. This structural evolution can be detected by Raman spectroscopy as a whole at different maturity levels no matter the dominant underlying reason. Previous studies evaluated these governing factors separately whereas here, we were able to see a combined effect of them via Raman spectra. Based on the conducted by Khatibi et al. [29], following conclusions can be drawn:

- Raman spectroscopy was able to detect molecular alterations in the organic matter as it undergoes thermal maturity.
- Raman mapping mode provided a way to compare different OM solid bitumen particles in microscale in a continuous format at varying maturity levels.
- HP although known as the best thermal maturity simulation method, does not necessarily follow natural maturation conditions. This was revealed by the cross-plot of Raman spectroscopy data versus conventional geochemistry results. The result can help to propose protocols and adjust experimental parameters for HP based on those cross-plots for more accurate thermal maturity progression steps in the lab.
- Raman spectroscopy can reflect heterogeneity of organic matter at microscale which can also represent expected heterogeneity in geochemical and geomechanical properties of the OM.
- It was found as maturity increases, heterogeneity of organic matter decreases, and Raman map of solid bitumen particle shows more uniform distribution of G-D band characteristics.

References

1. Abarghani A, Ostadhassan M, Gentzis T, Carvajal-Ortiz H, Bubach B (2018) Organofacies study of the Bakken source rock in North Dakota, USA, based on organic petrology and geochemistry. Int J Coal Geol 188:79–93
2. Amorin R (2009) Application of minimum curvature method to well-path calculations. PhD diss
3. Angulo S, Buatois LA (2012) Integrating depositional models, ichnology, and sequence stratigraphy in reservoir characterization: the middle member of the Devonian-Carboniferous Bakken formation of subsurface southeastern Saskatchewan revisited Bakken Formation Reservoirs, Saskatchewan, Canada. AAPG Bull 96(6):1017–1043
4. Bajc S, Ambles A, Largeau C, Derenne S, Vitorović D (2001) Precursor biostructures in kerogen matrix revealed by oxidative degradation: oxidation of kerogen from Estonian kukersite. Org Geochem 32:773–784
5. Behar F, Vandenbroucke M, Tang Y, Marquis F, Espitalié J (1997) Thermal cracking of kerogen in open and closed systems: determination of kinetic parameters and stoichiometric coefficients for oil and gas generation. Org Geochem 26:321–339
6. Beyssac O, Goffé B, Petitet J-P, Froigneux E, Moreau M, Rouzaud J-N (2003) On the characterization of disordered and heterogeneous carbonaceous materials by Raman spectroscopy. Spectrochim Acta Part A Mol Biomol Spectrosc 59(10):2267–2276
7. Carvajal-Ortiz H, Gentzis T (2015) Critical considerations when assessing hydrocarbon plays using Rock-Eval pyrolysis and organic petrology data: data quality revisited. Int J Coal Geol 152:113–122
8. Chen J, Xiao X (2014) Evolution of nanoporosity in organic-rich shales during thermal maturation. Fuel 129:173–181
9. Cheshire S, Craddock PR, Guangping X, Sauerer B, Pomerantz AE, McCormick D, Abdallah W (2017) Assessing thermal maturity beyond the reaches of vitrinite reflectance and Rock-Eval pyrolysis: a case study from the Silurian Qusaiba formation. Int J Coal Geol 180:29–45

10. Cramer B, Krooss BM, Littke R (1998) Modelling isotope fractionation during primary cracking of natural gas: a reaction kinetic approach. Chem Geol 149:235–250
11. Curtis ME, Cardott BJ, Sondergeld CH, Rai CS (2012) Development of organic porosity in the Woodford Shale with increasing thermal maturity. Int J Coal Geol 103:26–31
12. Dembicki H (2016) Practical petroleum geochemistry for exploration and production. Elsevier
13. Dieckmann V, Horsfield B, Schenk H (2000) Heating rate dependency of petroleum-forming reactions: implications for compositional kinetic predictions. Org Geochem 31:1333–1348
14. Dietrich AB (2015) The impact of organic matter on geomechanical properties and elastic anisotropy in the Vaca Muerta shale. PhD diss., Colorado School of Mines. Arthur Lakes Library
15. Epistalié J, Deroo G, Marquis F (1985) La pyrolyse Rock-Eval et ses applications. Deuxième partie. Revue de l'Institut français du Pétrole 40:755–784
16. Ferrari AC, Robertson J (2000) Interpretation of Raman spectra of disordered and amorphous carbon. Phys Rev B 61:14095
17. Garcette-Lepecq A, Derenne S, Largeau C, Bouloubassi I, Saliot A (2000) Origin and formation pathways of kerogen-like organic matter in recent sediments off the Danube delta (northwestern Black Sea). Org Geochem 31:1663–1683
18. Guedes A, Valentim B, Prieto A, Noronha F (2012) Raman spectroscopy of coal macerals and fluidized bed char morphotypes. Fuel 97:443–449
19. Hackley PC, Araujo CV, Borrego AG, Bouzinos A, Cardott BJ, Cook AC, Eble C, Flores D, Gentzis T, Gonçalves PA (2015) Standardization of reflectance measurements in dispersed organic matter: results of an exercise to improve interlaboratory agreement. Mar Pet Geol 59:22–34
20. Hackley PC, Cardott BJ (2016) Application of organic petrography in North American shale petroleum systems: a review. Int J Coal Geol 163:8–51
21. Hao F, Zou H, Gong Z, Yang S, Zeng Z (2007) Hierarchies of overpressure retardation of organic matter maturation: case studies from petroleum basins in China. AAPG Bull 91:1467–1498
22. Hu M, Cheng Z, Zhang M, Liu M, Song L, Zhang Y, Li J (2014) Effect of calcite, kaolinite, gypsum, and montmorillonite on Huadian oil shale kerogen pyrolysis. Energy Fuels 28:1860–1867
23. Jacob H (1989) Classification, structure, genesis and practical importance of natural solid oil bitumen ("migrabitumen"). Int J Coal Geol 11:65–79
24. Jehlička J, Urban O, Pokorný J (2003) Raman spectroscopy of carbon and solid bitumens in sedimentary and metamorphic rocks. Spectrochim Acta Part A Mol Biomol Spectrosc 59(10):2341–2352
25. Jubb AM, Botterell PJ, Birdwell JE, Burruss RC, Hackley PC, Valentine BJ, Hatcherian JJ, Wilson SA (2018) High microscale variability in Raman thermal maturity estimates from shale organic matter. Int J Coal Geol 199:1–9
26. Kadoura A, Nair AKN, Sun S (2016) Adsorption of carbon dioxide, methane, and their mixture by montmorillonite in the presence of water. Microporous Mesoporous Mater 225:331–341
27. Kelemen S, Fang H (2001) Maturity trends in Raman spectra from kerogen and coal. Energy Fuels 15:653–658
28. Khatibi S, Ostadhassan M, Aghajanpour A (2018) Raman spectroscopy: an analytical tool for evaluating organic matter. J Oil Gas Petrochem Sci 2:00007
29. Khatibi S, Ostadhassan M, Hackley P, Tuschel D, Abarghani A, Bubach B (2019) Understanding organic matter heterogeneity and maturation rate by Raman spectroscopy. Int J Coal Geol 206:46–64
30. Khatibi S, Ostadhassan M, Tuschel D, Gentzis T, Bubach B, Carvajal-Ortiz H (2018) Raman spectroscopy to study thermal maturity and elastic modulus of kerogen. Int J Coal Geol 185:103–118
31. Khatibi S, Ostadhassan M, Tuschel D, Gentzis T, Carvajal-Ortiz H (2018) Evaluating molecular evolution of Kerogen by Raman spectroscopy: correlation with optical microscopy and Rock-Eval pyrolysis. Energies 11(6):1–19

32. Khatibi S, Aghajanpour A, Ostadhassan M, Ghanbari E, Amirian E, Mohammed R (2018a) Evaluating the impact of mechanical properties of Kerogen on hydraulic fracturing of organic rich formations, SPE Canada unconventional resources conference. Society of Petroleum Engineers
33. Killops SD, Killops VJ (2013) Introduction to organic geochemistry. Wiley
34. Klentzman JL (2009) Geochemical controls on production in the Barnett Shale, Fort Worth Basin
35. Kong L, Ostadhassan MO, Sarout J, Ling K, Li C, Wang H (2018) Impact of thermal maturation on wave velocity in the Bakken Shale. Society of Petroleum Engineers, SPE Western Regional Meeting
36. Lewan M (1997) Experiments on the role of water in petroleum formation. Geochim Cosmochim Acta 61:3691–3723
37. Lewan MD, Roy S (2011) Role of water in hydrocarbon generation from Type-I kerogen in Mahogany oil shale of the Green River formation. Org Geochem 42:31–41
38. Lewan M, Ruble T (2002) Comparison of petroleum generation kinetics by isothermal hydrous and nonisothermal open-system pyrolysis. Org Geochem 33:1457–1475
39. Lewan M, Winters J, McDonald J (1979) Generation of oil-like pyrolyzates from organic-rich shales. Science 203:897–899
40. Lewan M (1993) Laboratory simulation of petroleum formation. Organic geochemistry. Springer, pp 419–442
41. Li C, Ostadhassan M, Guo S, Gentzis T, Kong L (2018) Application of PeakForce tapping mode of atomic force microscope to characterize nanomechanical properties of organic matter of the Bakken Shale. Fuel 233:894–910
42. Liang M, Wang Z, Zheng J, Li X, Wang X, Gao Z, Luo H, Li Z, Qian Y (2015) Hydrous pyrolysis of different kerogen types of source rock at high temperature-bulk results and biomarkers. J Petrol Sci Eng 125:209–217
43. Liu D, Xiao X, Tian H, Min Y, Zhou Q, Cheng P, Shen J (2013) Sample maturation calculated using Raman spectroscopic parameters for solid organics: methodology and geological applications. Chin Sci Bull 58(11):1285–1298
44. Liu K, Ostadhassan M, Kong L (2017) Pore structure heterogeneity in middle Bakken formation. In: 51st US rock mechanics/geomechanics symposium. American Rock Mechanics Association
45. Liu Y, Xiong Y, Li Y, Peng P (2017) Effect of thermal maturation on chemical structure and nanomechanical properties of solid bitumen. Mar Pet Geol
46. Lupoi JS, Fritz LP, Parris TM, Hackley PC, Solotky L, Eble CF, Schlaegle S (2017) Assessment of thermal maturity trends in Devonian-Mississippian source rocks using Raman spectroscopy: limitations of peak-fitting method. Front Energy Res 5:24
47. Lünsdorf NK (2016) Raman spectroscopy of dispersed vitrinite—methodical aspects and correlation with reflectance. Int J Coal Geol 153:75–86
48. Mackenzie A, McKenzie D (1983) Isomerization and aromatization of hydrocarbons in sedimentary basins formed by extension. Geol Mag 120:417–470
49. Mansuy L, Landais P, Ruau O (1995) Importance of the reacting medium in artificial maturation of a coal by confined pyrolysis. 1. Hydrocarbons and polar compounds. Energy Fuels 9:691–703
50. McTavish R (1998) The role of overpressure in the retardation of organic matter maturation. J Pet Geol 21:153–186
51. Michels R, Landais P (1994) Artificial coalification: comparison of confined pyrolysis and hydrous pyrolysis. Fuel 73:1691–1696
52. Michels R, Landais P, Torkelson B, Philp R (1995) Effects of effluents and water pressure on oil generation during confined pyrolysis and high-pressure hydrous pyrolysis. Geochim Cosmochim Acta 59:1589–1604
53. Monthioux M (1988) Expected mechanisms in nature and in confined-system pyrolysis. Fuel 67:843–847
54. Monthioux M, Landais P (1987) Evidence of free but trapped hydrocarbons in coals. Fuel 66:1703–1708

55. Monthioux M, Landais P, Monin J-C (1985) Comparison between natural and artificial maturation series of humic coals from the Mahakam delta, Indonesia. Org Geochem 8:275–292
56. Mumm AS, İnan S (2016) Microscale organic maturity determination of graptolites using Raman spectroscopy. Int J Coal Geol 162:96–107
57. Myers GA, Kehoe K, Hackley P (2017) Analysis of artificially matured Shales with confocal laser scanning Raman microscopy: applications to organic matter characterization. Unconventional Resources Technology Conference (URTEC)
58. Olson RK (2008) Cutting analyses, in coalbed methane and shale gas exploration strategies: Workshop for sorbed gas reservoir systems: AAPG course 15, AAPG annual convention, San Antonio, Texas, April 24–25, 35 p
59. Pan C, Geng A, Zhong N, Liu J, Yu L (2009) Kerogen pyrolysis in the presence and absence of water and minerals: amounts and compositions of bitumen and liquid hydrocarbons. Fuel 88:909–919
60. Pan C, Jiang L, Liu J, Zhang S, Zhu G (2010) The effects of calcite and montmorillonite on oil cracking in confined pyrolysis experiments. Org Geochem 41:611–626
61. Piani L, Robert F, Beyssac O, Binet L, Bourot-Denise M, Derenne S, Le Guillou C, Marrocchi Y, Mostefaoui S, Rouzaud JN (2012) Structure, composition, and location of organic matter in the enstatite chondrite Sahara 97096 (EH3). Meteorit Planet Sci 47, 8–29
62. Potgieter-Vermaak S, Maledi N, Wagner N, Van Heerden JHP, Van Grieken R, Potgieter JH (2011) Raman spectroscopy for the analysis of coal: a review. J Raman Spectrosc 42(2):123–129
63. Quirico E, Rouzaud J-N, Bonal L, Montagnac G (2005) Maturation grade of coals as revealed by Raman spectroscopy: progress and problems. Spectrochim Acta Part A Mol Biomol Spectrosc 61:2368–2377
64. Rullkötter J, Marzi R (1988) Natural and artificial maturation of biological markers in a Toarcian shale from northern Germany. Org Geochem 13:639–645
65. Sauerer B, Craddock PR, AlJohani MD, Alsamadony KL, Abdallah W (2017) Fast and accurate shale maturity determination by Raman spectroscopy measurement with minimal sample preparation. Int J Coal Geol 173:150–157
66. Schito A, Romano C, Corrado S, Grigo D, Poe B (2017) Diagenetic thermal evolution of organic matter by Raman spectroscopy. Org Geochem 106:57–67
67. Schmidt JS, Hinrichs R, Araujo CV (2017) Maturity estimation of phytoclasts in strew mounts by micro-Raman spectroscopy. Int J Coal Geol 173:1–8
68. Schrader B (2008) Infrared and Raman spectroscopy: methods and applications. Wiley
69. Schumacher M, Christl I, Scheinost AC, Jacobsen C, Kretzschmar R (2005) Chemical heterogeneity of organic soil colloids investigated by scanning transmission X-ray microscopy and C-1s NEXAFS microspectroscopy. Environ Sci Technol 39:9094–9100
70. Spötl C, Houseknecht DW, Jaques RC (1998) Kerogen maturation and incipient graphitization of hydrocarbon source rocks in the Arkoma Basin, Oklahoma and Arkansas: a combined petrographic and Raman spectrometric study. Org Geochem 28:535–542
71. Steptoe A (2012) Petrofacies and depositional systems of the Bakken Formation in the Williston Basin, North Dakota. Masters Abs Int 51(03)
72. Tegelaar EW, Noble RA (1994) Kinetics of hydrocarbon generation as a function of the molecular structure of kerogen as revealed by pyrolysis-gas chromatography. Org Geochem 22:543–574
73. Tian H, Xiao X, Wilkins R, Li X, Gan H (2007) Gas sources of the YN2 gas pool in the Tarim Basin—evidence from gas generation and methane carbon isotope fractionation kinetics of source rocks and crude oils. Mar Pet Geol 24:29–41
74. Tissot BP, Welte DH (1984) Diagenesis, catagenesis and metagenesis of organic matter. Petroleum formation and occurrence. Springer, pp 69–73
75. Tuinstra F, Lo Koenig J (1970) Raman spectrum of graphite. J Chem Phys 53(3):1126–1130
76. Tyson RV (1995) Abundance of organic matter in sediments: TOC, hydrodynamic equivalence, dilution and flux effects. Sedimentary organic matter. Springer, pp 81–118
77. Waples DW (1981) Organic geochemistry for exploration geologists. Burgess Pub. Co.

78. Wu LM, Zhou CH, Keeling J, Tong DS, Yu WH (2012) Towards an understanding of the role of clay minerals in crude oil formation, migration and accumulation. Earth Sci Rev 115:373–386

79. Yang J, Hatcherian J, Hackley PC, Pomerantz AE (2017) Nanoscale geochemical and geomechanical characterization of organic matter in shale. Nat Commun 8:2179

80. Yang C-S, Kao S-P, Lee F-B, Hung P-S (2004) Twelve different interpolation methods: a case study of Surfer 8.0. Proceedings of the XXth ISPRS congress 35:778–785

81. Zargari S (2015) Effect of thermal maturity on nanomechanical properties and porosity in organic rich shales (a Bakken shale case study). PhD diss., Colorado School of Mines. Arthur Lakes Library

82. Zhang L, Buatois LA (2016) Sedimentology, ichnology and sequence stratigraphy of the Upper Devonian-Lower Mississippian Bakken formation in eastern Saskatchewan. Bull Can Pet Geol 64(3):415–437

83. Zhou Q, Xiao X, Pan L, Tian H (2014) The relationship between micro-Raman spectral parameters and reflectance of solid bitumen. Int J Coal Geol 121:19–25

84. Zhu G, Zhang S, Su J, Zhang B, Yang H, Zhu Y, Gu L (2013) Alteration and multi-stage accumulation of oil and gas in the Ordovician of the Tabei Uplift, Tarim Basin, NW China: implications for genetic origin of the diverse hydrocarbons. Mar Pet Geol 46:234–250

Chapter 5
Backtracking to Parent Maceral from Produced Bitumen with Raman Spectroscopy

Abstract In order to assess a source rock for economical exploitation purposes, many parameters should be considered; regarding the geochemical aspects the most important ones are the amount of organic matter (OM) and its quality. Quality refers to the thermal maturity level and the type of the OM from which it was formed. The origin of the OM affects the ability of the deposited OM between sediments to generate oil, gas or both with particular potential after going through thermal maturation. Vitrinite reflectance and programmed pyrolysis (for instance Rock–Eval) are common methods for evaluating the thermal maturity of the OM and its potential to generate petroleum, but do not provide us with an answers to what extent solid bitumen is oil-prone or gas-prone since they are bulk geochemical methods. In the present study, Raman spectroscopy (RS), as a powerful tool for studying carbonaceous materials and organic matter, was conducted on shale and coal samples and their individual macerals to show the potential of this technique in kerogen typing and revealing the parent maceral of the examined bitumen. The proposed methodology by exhibiting the chemical structure of different organic matters as a major secondary product in unconventional reservoirs can also detect the behavior of solid bitumen and its hydrocarbon production potential for more accurate petroleum system evaluation.

Keywords Raman spectroscopy · Maturity level · Kerogen type · Solid bitumen · Maceral

5.1 Introduction

Kerogen is a macromolecule (geopolymer) that forms the insoluble portion of the dispersed organic matter (OM) in source rocks and generates oil and natural gas through thermal maturation [42, 92]. The capability of the kerogen to generate oil or gas is dependent upon the maceral composition [86, 94] and its burial conditions (pressure and temperature), however, the effects of local parameters should not be ignored either [1, 71]. Thermal maturity level is traditionally measured by reflectance methods such as vitrinite, solid bitumen, or Zooclasts (graptolites, chitinozoans, and

© The Author(s), under exclusive license to Springer Nature Switzerland AG 2024   87
M. Ostadhassan and B. Hazra, *Advanced Methods in Petroleum Geochemistry*,
SpringerBriefs in Petroleum Geoscience & Engineering,
https://doi.org/10.1007/978-3-031-44405-0_5

scolecodonts) reflectance as the most reliable approaches. The biogenic origin of the kerogen also known as the parent maceral is usually represented by the particulate organic matter and bitumen that is distinguished in rock samples as phytogenic organic substance with distinct chemical/physical properties, through microscopic examination (transmitted and reflected white, and ultraviolet lights) [22, 23, 43, 72, 79, 81, 82, 84, 85] (ASTM D7708-14, 2014). Other methods include infrared spectroscopy (through the various ratios of aromatic/aliphatic compounds [2, 56]), programmed pyrolysis (e.g., Rock Eval derived T_{max}), liquid chromatography, gas chromatography, gas chromatography coupled with mass spectrometry, etc., could provide some information about its thermal maturity too [5, 54, 86]. Kerogen typing is not only useful to detect the origin and depositional environment (lacustrine, marine and terrestrial) of recognized organic matter, but it also helps to develop models in regard to the OM petrophysical characteristics such as the evolving pore spaces at different thermal maturities [39, 58]. Ultimately, the combination of these studies, will enable more precise petroleum system evaluation, petrophysical characterization and finally reserve estimation in organic-rich shale plays.

Based on extensive organic petrography studies of source rock specimens, three major maceral groups of kerogens (Table 5.1) are identified considering their unique morphologies and grayness in reflected light [16, 73]:

- The liptinite group (Type I and II kerogen) includes primarily algal material or amorphous organic matter derived from algal or bacterial precursors,
- the vitrinite group (Type III kerogen) refers to organic matter that is derived from the woody tissue of post-Silurian vascular plants,
- and the inertinite group (Type IV kerogen) includes macerals that have experienced combustion, oxidation, or desiccation [32, 70].

In addition to these three major maceral groups, zooclasts and secondary products (solid bitumen) are also considered in the present study. Zooclasts are organic matter

Table 5.1 Maceral groups classification of The Society for organic petrology/international commission for coal and organic petrology (TSOP/ICCP)

Maceral group	Kerogen type	Grayness	Some of the macerals in the group
Liptinite	I and II	Dark gray	Alginite, sporinite, cutinite, suberinite, resinite, liptodetrinite
Vitrinite	III	Medium to light gray	Collotelinite, vitrodetrinite, telinite
Inertinite	IV	White and can be very bright	Fusinite, macrinite, micrinite, funginite
Zooclast	–	Medium to light gray	Graptolite, scolecodont
Secondary products	–	Medium to light gray	Solid bitumen

Modified from: [25, 26, 28]

Zooclast and secondary products are not categorized in any kerogen type with reflecting surfaces that have a similar grayness to vitrinite or solid bitumen that can be distinguished based on their unique morphologies [13]. Secondary products are the consequence of OM transformation and are usually referred to as solid bitumen or other by-products. Secondary products like solid bitumen are an important thermal maturity indicator when vitrinite is scarce/absent [7, 8, 44, 60, 83]. Although solid bitumen is not considered as an independent kerogen type, it becomes the predominant organic matter at higher thermal maturities in samples within the late oil to dry gas window [32, 53, 62, 74].

Kerogen constituents classifications are primarily based on their appearance and preservation states, using transmitted and reflected white light, and investigating fluorescence emission properties utilizing the ultraviolet light. These methods examine the kerogen morphology and emphasis is made on the botanical source, and/or preservation states [86, 90].

Under transmitted white light, kerogen can be divided into: Phytoclast group (fragments of tissues derived from higher plants or fungi); Palynomorph group (organic-walled microfossils composed of entirely unmineralized proteinaceous material, such as: chitinozoans, spores and pollen, prasinophytes, acritarchs, and dinoflagellates); Zooclast Group (faunal relics such as chitinozoans, graptolites, scolecodonts, and conodonts); and, Amorphous Group (Amorphous Organic Matter) [65, 90, 93]. However, "Solid bitumen" is categorized as another group of organic matter occurring in sedimentary rocks. It is formed after the thermal decomposition of kerogen during the incipient-oil generation stage, then decomposes to oil in the primary-oil generation stage, while in the post-oil generation phase the solid bitumen is decomposed to gas or pyrobitumen [86] (Lewan 1993).

Incident light microscopy, optical reflectance and programmed pyrolysis are conventional methods for evaluating maturity level of the organic matter and a pathway to identify kerogen type [25, 26, 32]. However, these methods become challenging to be utilized on small volumes of OM or when a specific maceral cannot be recognized in the samples [40, 58]. Additionally, programmed pyrolysis is a bulk analyzer that merely yields general information about the dominant kerogen type within the samples, and in overmature samples it is difficult to recognize initial maceral composition from programmed pyrolysis. Thus, programmed pyrolysis is not sensitive to any maceral specific information about the individual kerogen types within the sample, which means the unique impact of each component to the final result is completely ignored [30]. Thus, to overcome this shortcoming, organic petrography should be utilized to delineate existing types of kerogen macerals in a sample for detailed source rock evaluation. However, this method is exhaustive and requires significant training and expertise. Furthermore, petrographic approaches cannot state whether solid bitumen, as an individual particle, is oil-prone or gas prone and, more in detail, what is the original maceral that produced the solid bitumen. Therefore, employing more elaborate analytical methods such as Raman spectroscopy (RS) can contribute towards a more accurate maceral characterization at smaller scales for individual OM particles [29, 30, 61, 63, 64, 67, 96, 98]. Employing RS has also shown potential to detect detailed composition of different OM types

[58], and the impact of each component on hydrocarbon production by understanding their rate of maturation and their oil and gas proneness [29–31, 66, 67, 98].

RS operates based on molecular vibrations to reveal the molecular structure of substances, such as the organic matter [9, 10, 34, 50, 76] and inorganic minerals [6, 28, 36, 89]. In this method, after illuminating the sample surface with laser light, a corresponding spectrum will be generated which comprises some bands as finger-prints that refers to specific chemical structures. Raman spectrometers are easy to use and are commonly equipped with an optical microscope to identify and then analyze specific components in source rocks, including individual recognized macerals or solid bitumen [29–31, 67, 99]. Taking advantage of RS has helped researchers to better understand thermal maturity trends of different organic matters [21, 31, 37, 45, 84, 86, 89], and also provides information about their H/C ratio [19] and geochemical and even geomechanical properties [49–51].

Based on what is laid out above, Khatibi et al. [48] employed Raman spectra signals to investigate the potential application of this method on kerogen typing or backtracking to the parent maceral that has produced the studied kerogen particle. It was done based on revealing the chemical structural information of different organic matter including macerals from three types of kerogen and solid bitumen and relating them statistically. Ultimately, based on relationships, the proposed method was applied to solid bitumen solely to investigate if RS can reveal the originating kerogen type that has led to the solid bitumen through thermal maturation. Their study was an attempt for more accurate geochemical analysis of source rocks and shale plays for cost-effective exploitation and exploration purposes. They collected samples from Penn State Coal Bank, Green River, Eagle Ford, Woodford, and Bakken Formations (Fig. 5.1) to make sure a wide range of maturities and identified kerogen types are included, as shown in Fig. 5.2. They retrieved coal samples were retrieved from Penn State Coal bank samples as a set of vitrinite-rich coals of increasing rank from low to high rank A. The Green River Formation is an Eocene aged geologic formation that has the sedimentation in a group of intermountain lakes in Colorado, Wyoming, and Utah, USA. Samples were retrieved from depth of 12,200–12,280 ft with maturies from $VR_{O\text{-}Eq} \sim 1.56$ to 1.59 [30]. The Eagle Ford Shale is a Late Cretaceous sedimentary rock located in Texas, USA. The Eagle Ford is composed of organic matter-rich fossiliferous marine shales and marls with interbedded thin limestones. Samples were retrieved from depth of 7200–9100 ft with maturies from $VR_{O\text{-}Eq} \sim 0.95$ to 1.1. The Woodford Shale covers the state of Oklahoma, USA and is mostly Late Devonian in age with the uppermost part as Early Mississippian [32]. Samples were retrieved from depth of 12,200 to 12,280 ft with maturies from $VR_{O\text{-}Eq} \sim 0.55$ to 4.3. The Bakken Formation is a rock unit from the Late Devonian to Early Mississippian age in the Williston Basin, underlying parts of Montana, North Dakota in USA and Saskatchewan and Manitoba in Canada. This formation consists of three members: lower shale, middle dolomite, and upper shale. The shales were deposited in relatively deep anoxic marine conditions and samples were retrieved from depth of 5438–11,199 ft with maturies from $VR_{O\text{-}Eq} \sim 0.38$ to 0.92 [1].

Fig. 5.1 The map of where the samples were retrieved from, showing major shale plays in North America

5.2 Raman Spectroscopy and Maturity

The molecular structure of OM has a strong correlation with its maturity level from the functional groups variation stand point. Previous studies have documented systematic changes in the structure of the OM by RS (band positions, band separation, FWHM etc.) as a function of maturity advancement (e.g., [19, 45, 49–51, 80]). When kerogen advances towards higher maturities, its aromaticity will increase and, in this process, the macromolecule breaks down and loses its O, N and S content [50, 88], which can be detected by Raman response. Figure 5.3 depicts the correlation between separation of two major Raman bands (G-D) versus equivalent vitrinite (VR_{O-Eq}) maturity of the samples retrieved from different formations. The increase of the band separation with maturity is attributed to the shift in the D band towards lower wavenumbers with a minor change in the G band towards higher wavenumbers. Moreover, at lower maturity levels, band separation changes at a higher rate, however, towards the higher levels of thermal maturity, the rate decreases and gradually the separation of the G-D bands tapers off. This can be interpreted as advancement to a more homogenous OM; This happens because by increasing maturity H/C and O/C ratios decrease and the final composition of the OM approaches to the pure carbon (graphite) which can be reflected by minor changes in band separation [91].

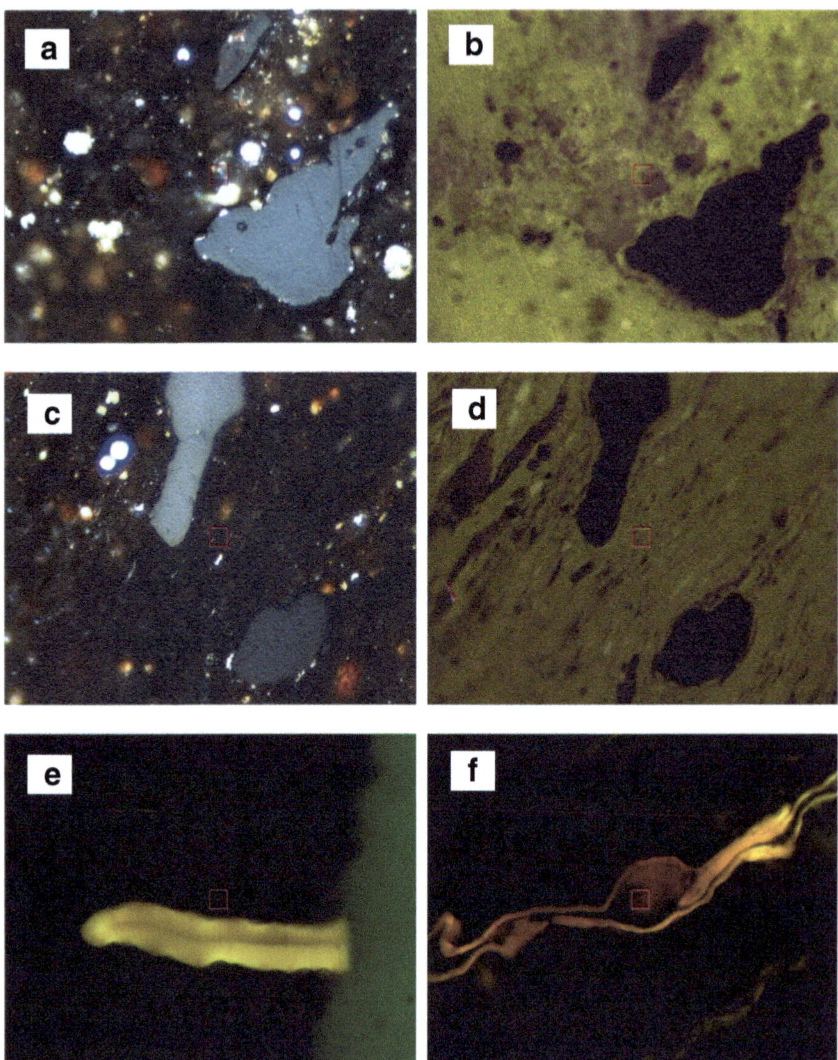

Fig. 5.2 **a** Inertinite particles (gray) in a matrix of amorphous organic matter (AOM); **b** the same view under UV light. Inertinite particles have no fluorescence, the AOM is recognizable with yellowish green fluorescence; **c** inertinite particles (gray) in a matrix of lamellar algal matter and AOM; **d** the same view under UV light. Lamellar algal matter is recognizable by dull yellowish green fluorescence. Photomicrographs **a–d** represent type I and IV kerogens; **e** Alginite (*Tasmanites*) with dull golden yellow fluorescence under UV light, the central suture line is clearly visible; **f** *Leiosphaeridia alginite* with golden-yellow fluorescence. Photomicrographs **e–f** represent type II kerogen; **g** elongated low-reflecting solid bitumen (gray), (SBRo,ran = 0.27%); **g** solid bitumen, (SBRo,ran = 0.40%); photomicrographs **a–d** and **e–h** were taken from the samples of the green river formation and the Bakken formation, respectively, using a 50× oil immersion objective. The red square in the middle of each image is a scale of 5 μm of each side. **i, j** Collotelinite and fusinite samples, images are extracted from Guedes et al. [30]

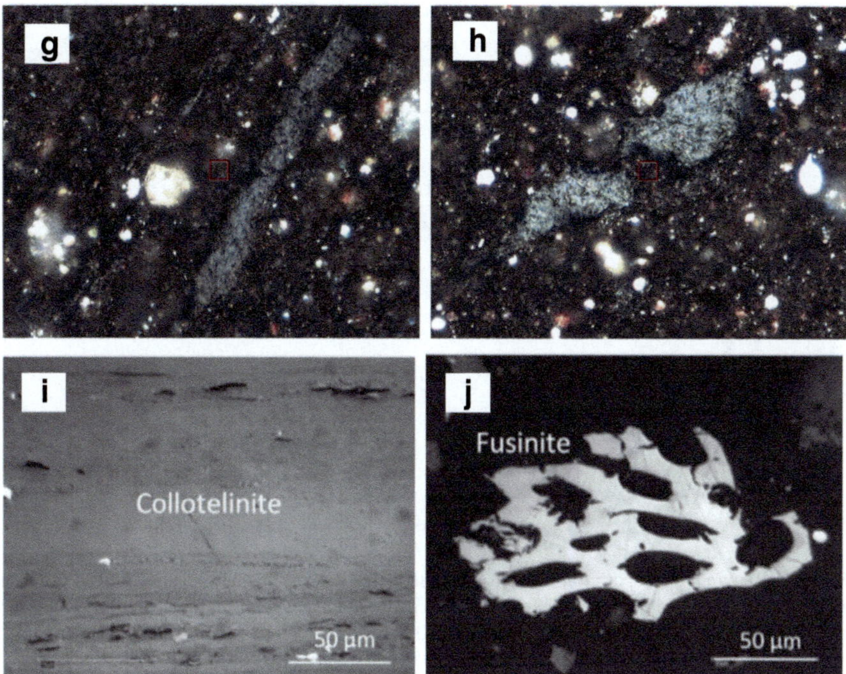

Fig. 5.2 (continued)

Khatibi et al. [48] also plotted each maceral's reflectance versus G-D band separation to better delineate the existance of different chemical information in their samples, Fig. 5.4. Since each OM in this graph should own different maceral origin and consequently different physico-chemical properties, they demonstrate various rates of band separation versus maturity as they progress towards aromatization [67]. This phenomenon is represented by the slope of the curves that is fitted through the data for each group of macerals. As seen in Fig. 5.4, fusinite and macrinite (Type IV kerogen) show similar trends in the curves. This similarity can be related to the fact that fundamentally both are derived from the same origin (plants) and resulting from the decay path before or during peatification [30, 41, 78]. Collotelinite and another (undisclosed) Type III kerogen are also demonstrating similar trends. Graptolites (zooclasts) and Type II (unknown maceral) kerogen, even though from different origins, have similar slopes, which is in agreement with the Raman studies of Bustin et al. [12] and Mumm and İnan [67].

At the same level of thermal maturity, general existing fraction of aromatic carbon in kerogen has the order of: Type IV > Type III > Type II [11, 56, 70, 88], which is exactly opposite the slopes seen in Fig. 5.4. It can be described that the rate of maturation of different organic matters would differ and resulting in a varying rate of band separation. This varying maturity rate is equivalent to different rate (which corresponds to different slopes in the figures) of advancement towards higher

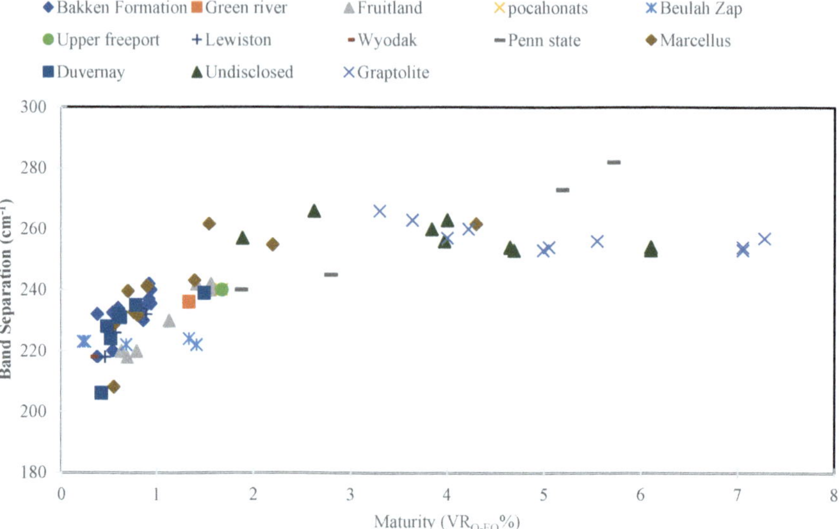

Fig. 5.3 Raman band separation (G-D) versus maturity (VR$_{O-Eq}$%) for 13 samples from different geologic regions including the Bakken Formation, Green River, Penn State, Marcellus, Duvernay and undisclosed Formations. Modified from Khatibi et al. [51]. The data in this figure is a combination of data collected by Khatibi et al. [48] and others [45, 51, 75, 80]

aromaticity levels in each kerogen type or OM particle. Furthermore, researchers have also documented that at the same maturity level, reflectance of kerogens types is in the order of Type IV > Type III > Type II prior to converting the reflectance values to the equivalent vitrinite reflectance [14, 17, 97]. This has been inferred to be due to different aromaticity levels of various kerogen types and OM [33, 59, 60]. Therefore, slopes of different organic matter in Fig. 5.4 represent the macromolecular aromatic components of OM which is also confirming their optical characteristics.

5.3 Raman Spectroscopy and Kerogen Typing

Each kerogen type comprises from a set of macerals that share almost similar physical and chemical properties [35], which make them more closely related compared to macerals from other kerogen types [42]. Based on this idea, RS as a tool which provides insight to study molecular structures through vibrational methods (a physical characteristic related to chemical structure) should have the potential in separating various types of kerogen and delineating the pathway of OM overall evolution.

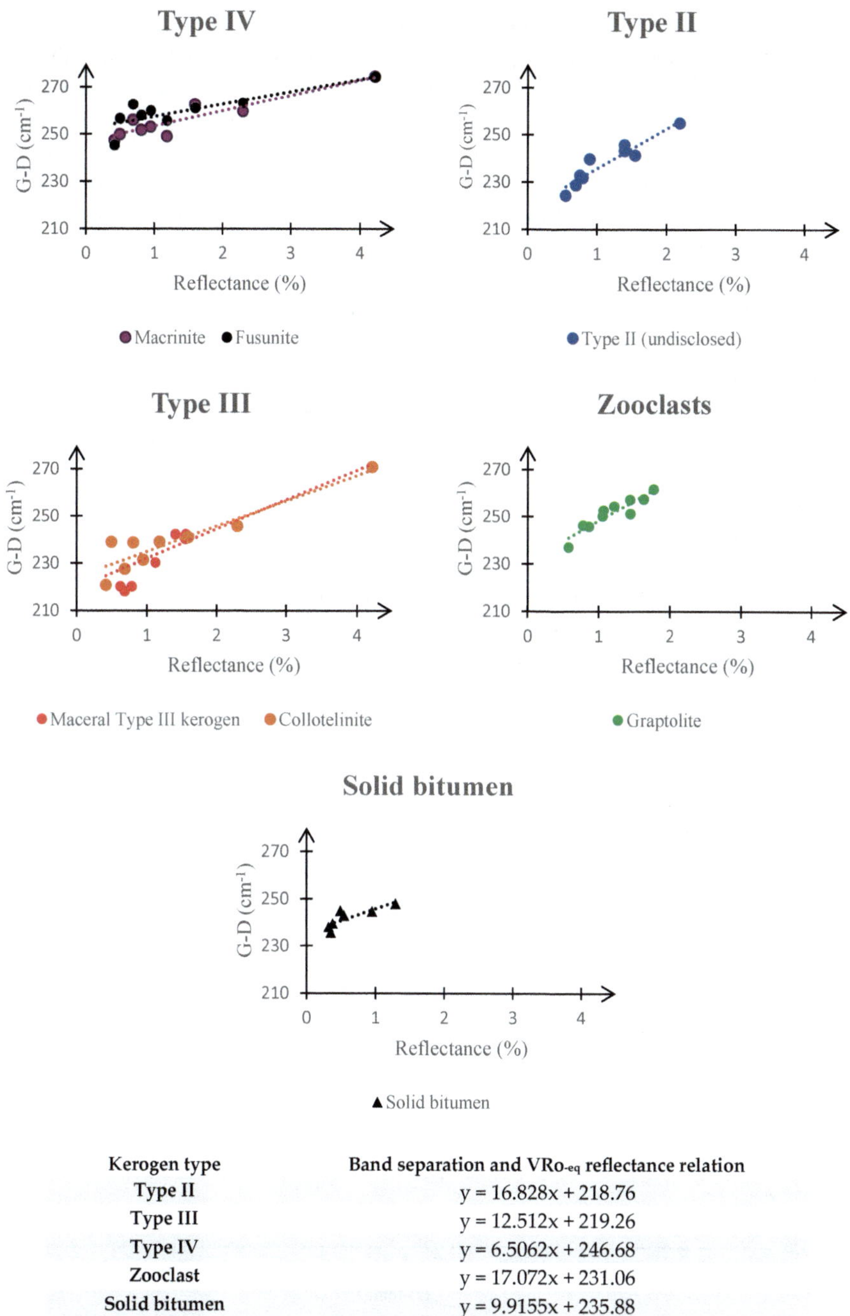

Kerogen type	Band separation and VRo-eq reflectance relation
Type II	$y = 16.828x + 218.76$
Type III	$y = 12.512x + 219.26$
Type IV	$y = 6.5062x + 246.68$
Zooclast	$y = 17.072x + 231.06$
Solid bitumen	$y = 9.9155x + 235.88$

Fig. 5.4 Comparison of band separation and reflectance of several OM types. Data were extracted by Khatibi et al. [48] from Kelemen and Fang [45], Guedes et al. [30], Liu et al. [57], Mumm and İnan [67]

To do so, Khatibi et al. [48] plotted different existing bands (wavenumbers) against one another from the studied samples and realized among all, G (vibration within the aromatic ring) band versus D5 (vibrations originating from aliphatic hydrocarbon chains) displayed a good distinction of different kerogen types, Fig. 5.5, which was in accordance with the study of Liu et al. [58]. Based on the position of each kerogen type in this diagram, it can be inferred that sulfur (S) content of the studied OMs is playing a key role in separation of various kerogen types. Table 5.2 explains the average elemental composition of immature kerogens reported by Dembicki [18]. It can be found that type II kerogen has the highest sulfur content, type I the least and type III in between [3, 87] which follows the trend seen in Fig. 5.5. Lacustrine environments yield in kerogens which are relatively depleted in S, while marine environments based on anoxic/euxinic conditions and mineralogy result in kerogens with varying amounts of organic S [46]. Type II-S kerogen contains higher amount of S (8–14 wt%) compared to other kerogen types and is derived from autochthonous OM in highly reducing conditions in marine environment which is mostly associated with upwelling conditions [18, 68]. As seen in Fig. 5.5, type II-S is located at the far right region of this graph representing the OM with the highest sulfur content.

D5 band has been uniquely related to vibrations originating from aliphatic hydrocarbon chains [19]. Li et al. [55] utilized light-element electron microprobe (EMP) and micro-ATR–FTIR to study macerals in bituminous coals with variable organic sulfur content from different geographic locations. This study argues that the higher organic S content in the maceral appears to be accompanied by a greater proportion of aliphatic functional groups, as a result of replacement of some of the oxygen (O) within the maceral in ring structures with sulfur (S). This can explain the trend that

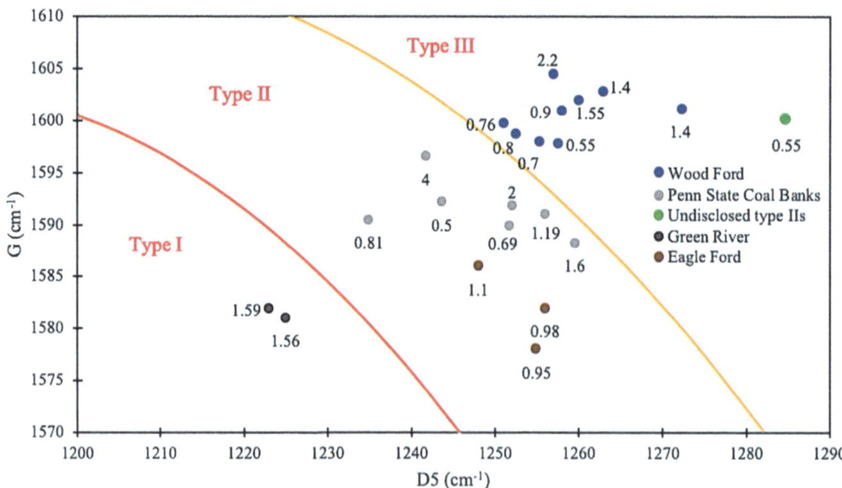

Fig. 5.5 G versus D5 is important for kerogen typing. As seen, different types of kerogen are located on this graph based on what is expected from their sulfur content (Type II > Type III > Type I) (numbers reflect thermal maturity of the samples)

Table 5.2 Average elemental composition of C, H, O, S and N in kerogen

	%Carbon	%Hydrogen	%Oxygen	%Sulfur	%Nitrogen
Type I	80	10.9	4.9	1.3	1.6
Type II	68.9	7.3	6.6	10.6	1.5
Type III	56.2	4.4	27.8	2.4	1.6

After Dembicki [18]

is observed in Fig. 5.5, where kerogens with higher sulfur content (corresponds to greater proportion of aliphatic functional groups) are exhibiting higher D5 values. Moreover, Kelemen et al. [46] used X-ray photon spectroscopy (XPS) and sulfur X-ray absorption near edge structure (S-XANES) to analyze a wide range of organic matter types and maturities. They detected that when the amount of aromatic carbon is increased, aliphatic sulfur declines for all kerogen types. This idea can also corroborate the distribution trend of macerals in the changing format from upper left to the lower right regions in this plot. G band is purely of aromatic character [19] and the shift of G band towards higher wave numbers by increasing aromaticity has also been observed by others [45, 67]. Thus, when G band wavenumber is increased (shifting towards upper left region of the graph for each kerogen type that are separated) D5 wavenumber would decrease which can be associated with a decrease in aliphatics (sulfur).

As mentioned earlier, solid bitumen is a secondary product which is not categorized under any kerogen type, while it is a product of any type of kerogen (I, II or III) conversion through thermal maturity or petroleum de-asphalting process [62]. Therefore, it is expected that any solid bitumen based on the parent maceral to display somehow similar physicochemical properties to a particular type of kerogen that it is basically originating from. Gas/oil proneness of solid bitumen which explains its hydrocarbon generation potential, and its role in thermal maturity evaluation make solid bitumen an important particle in unconventional organic-liquid-rich shale plays [24, 47, 52, 62, 77]. However, organic petrography is inadequate for accurately detecting gas/oil proneness and incapable of relating the identified solid bitumen particle to its parent maceral. In order to investigate if the results from this study that was performed on macerals can be extended to solid bitumen and reveal the kerogen type that specific solid bitumen is originating from, Khatibi et al. [48] utilized samples from the Bakken and Eagle Ford Shale Formations and examined solid bitumen particles that were identified to acquire their Raman spectra. G and D5 bands from studied solid bitumen particles are overlain on the same plot as in Fig. 5.5 and presented in Fig. 5.6. It is observed, two separate solid bitumen populations are delineated which is indicating properties similar to two different types of kerogen. This can be supported by the fact that solid bitumen is a product of OM transformation.

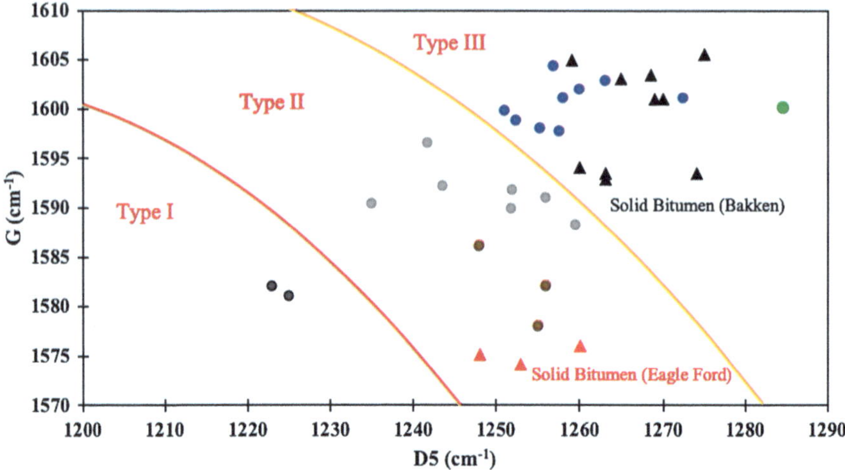

Fig. 5.6 G versus D5 for kerogen typing. Solid bitumen samples are also shown in this figure. As seen, solid bitumen populations can show different characters. Data points are the same as Fig. 5.5, except red and black triangles that respectively represent solid bitumen from the Eagle Ford and Bakken

Figure 5.7 is the pseudo-Van Krevelen diagram from bulk analysis of shale samples from the Bakken Formation pointing to the type II as the dominant kerogen type in the samples that their solid bitumen were examined with Raman. Bulk analysis such as programmed pyrolysis will describe relative abundance of kerogen types. However, considering solid bitumen as the predominant organic matter in late oil to dry gas window [32, 53, 62, 74], it is reasonable to assume that programmed pyrolysis is reflecting properties of the solid bitumen relevant to type II kerogen, mostly, from which it has been originating. While, solid bitumen from Eagle Ford is presenting properties similar to type III kerogen enforcing the separation of the data between these two studied solid bitumen particles. Therefore, we can propose utilizing RS is able to detect the parent kerogen of the existing solid bitumen in the samples.

The results can also explain the difference in the trends between some of the major equations that have been introduced to convert solid bitumen reflectance to $VR_{O\text{-}Eq}$ in the literature. These equations are developed from measurements on solid bitumen that are the product of various types of kerogen conversion through thermal maturity and therefore discrepancies should be expected.

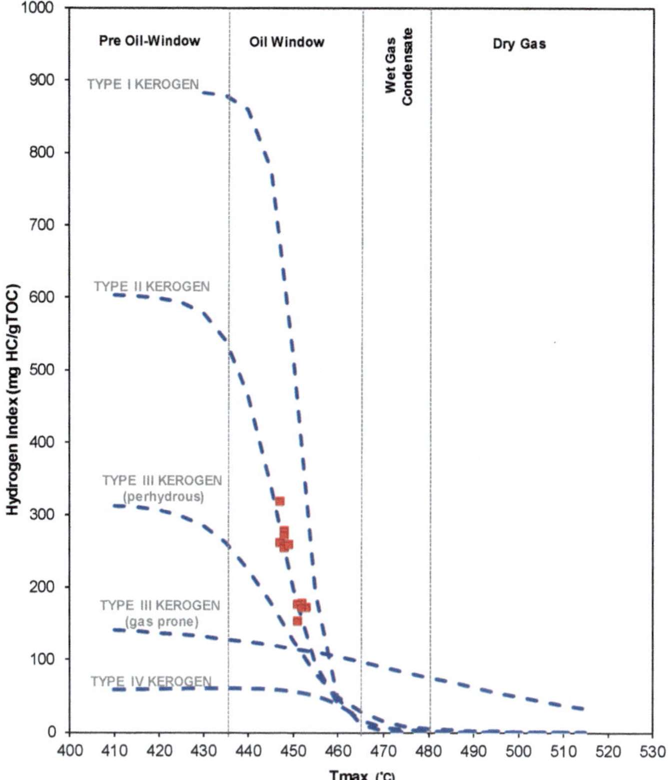

Fig. 5.7 Hydrogen index versus T_{max} for kerogen typing using Rock–Eval parameters for the Bakken samples

5.4 Conclusion

Bulk analyses such as programmed pyrolysis on OM can provide general information without considering the impact of each component to the result. In the study by Khatibi et al. [48], use of RS as a well-known tool for studying carbonaceous materials, the potential of kerogen typing using Raman signals was investigated which can benefit from being applicable directly to embedded OM particles in sediments (no need to kerogen isolation) in a fast and accurate way without high labor cost.

Based on the results of Khatibi et al. [48] it was interpreted that plotting peak positon of G band versus D5 will distinguish different kerogen types based on their aromaticity level and sulfur content. In the next step the proposed method was performed on solid bitumen samples to support the idea of different behaviors and multiple populations of solid bitumen in shale rocks which can lead to a new path on further studies on solid bitumen as one of the key component in shale plays and source rocks as hydrocarbon generator, thermal maturity indicator, hydrocarbon storage and migration.

References

1. Abarghani A, Ostadhassan M, Gentzis T, Carvajal-Ortiz H, Ocubalidet S, Bubach B, Mann M, Hou X (2019) Correlating Rock-Eval™ Tmax with bitumen reflectance from organic petrology in the Bakken formation. Int J Coal Geol 205:87–104

2. Abarghani A, Gentzis T, Shokouhimehr M, Liu B, Ostadhassan M (2020) Chemical heterogeneity of organic matter at nanoscale by AFM-based IR spectroscopy. Fuel 261:116454

3. Altun NE, Hicyilmaz C, Hwang J-Y, Suat Bagci A, Kok MV (2006) Oil shales in the world and Turkey; reserves, current situation and future prospects: a review. Oil Shale 23(3):211–228

4. Baludikay BK, François C, Sforna MC, Beghin J, Cornet Y, Storme J-Y, Fagel N et al (2018) Raman microspectroscopy, bitumen reflectance and illite crystallinity scale: comparison of different geothermometry methods on fossiliferous Proterozoic sedimentary basins (DR Congo, Mauritania and Australia). Int J Coal Geol 191:80–94

5. Behar F, Beaumont V, De Penteado HLB (2001) Rock-Eval 6 technology: performances and developments. Oil Gas Sci Technol 56(2):111–134

6. Bersani D, Lottici PP (2016) Raman spectroscopy of minerals and mineral pigments in archaeometry. J Raman Spectrosc 47(5):499–530

7. Bertrand R, Malo M (2001) Source rock analysis, thermal maturation and hydrocarbon generation in Siluro-Devonian rocks of the Gaspe Belt basin, Canada. Bull Can Pet Geol 49(2):238–261

8. Bertrand R (1993) Standardization of solid bitumen reflectance to vitrinite in some Paleozoic sequences of Canada. Energy Sour 15(2):269–287

9. Beyssac O, Goffé B, Chopin C, Rouzaud JN (2002) Raman spectra of carbonaceous material in metasediments: a new geothermometer. J Metamorph Geol 20(9):859–871

10. Beyssac O, Goffé B, Petitet J-P, Froigneux E, Moreau M, Rouzaud J-N (2003) On the characterization of disordered and heterogeneous carbonaceous materials by Raman spectroscopy. Spectrochim Acta Part A Mol Biomol Spectrosc 59(10):2267–2276

11. Borrego AG, Hagemann HW, Prado JG, Guillén MD, Blanco CG (1996) Comparative petrographic and geochemical study of the Puertollano oil shale kerogens. Org Geochem 24(3):309–321

12. Bustin RM, Link C, Goodarzi F (1989) Optical properties and chemistry of graptolite periderm following laboratory simulated maturation. Org Geochem 14(4):355–364

13. Cardott BJ, Kidwai MA (1991) Graptolite reflectance as a potential thermal-maturation indicator. Okla Geol Surv Circular 92:203–209

14. Cardott BJ (2012) Thermal maturity of Woodford Shale gas and oil plays, Oklahoma, USA. Int J Coal Geol 103:109–119

15. Cesare B, Maineri C (1999) Fluid-present anatexis of metapelites at El Joyazo (SE Spain): constraints from Raman spectroscopy of graphite. Contrib Miner Petrol 135(1):41–52

16. Chaudhuri SN (2016) Coal Macerals. Encycl Miner Energy Policy:1–5

17. Crelling JC, Dutcher RR (1980) Principles and applications of coal petrology: short course notes. No. 8. Soc Econ Paleontol Mineral

18. Dembicki H (2016) Practical petroleum geochemistry for exploration and production. Elsevier

19. Ferralis N, Matys ED, Knoll AH, Hallmann C, Summons RE (2016) Rapid, direct and non-destructive assessment of fossil organic matter via microRaman spectroscopy. Carbon 108:440–449

20. Ferrari AC, Robertson J (2001) Resonant Raman spectroscopy of disordered, amorphous, and diamond like carbon. Phys Rev B 64(7):075414

21. Ferrari AC, Robertson J (2000) Interpretation of Raman spectra of disordered and amorphous carbon. Phys Rev B 61(20):14095

22. For Coal, International Committee (2001) The new inertinite classification (ICCP System 1994). Fuel 80(4):459–471

23. For Coal, International Committee (1998) The new vitrinite classification (ICCP System 1994). Fuel 77(5):349–358

24. Gentzis T, Goodarzi F (1990) A review of the use of bitumen reflectance in hydrocarbon exploration with examples from Melville Island, Arctic Canada. Rocky Mountain Section (SEPM)

25. Gentzis T, Carvajal-Ortiz H, Tahoun S, Li C, Ostadhassan M, Xie H, Filho JGM (2017) A multi-component approach to study the source-rock potential of the Bakken Shale in North Dakota, USA, using organic petrology, Rock-Eval pyrolysis, palynofacies. LmPy-GCMSMS geochemistry, and NMR spectroscopy

26. Gentzis T, Carvajal-Ortiz H, Ocubalidet SG, Wawak B (2017) Organic petrology characteristics of selected shale oil and shale gas reservoirs in the USA: examples from "the magnificent nine", Geology: current and future developments: the role of organic petrology in the exploration of conventional and unconventional hydrocarbon systems 1:131–168

27. Gorbanenko OO, Ligouis B (2014) Changes in optical properties of liptinite macerals from early mature to post mature stage in Posidonia Shale (Lower Toarcian, NW Germany). Int J Coal Geol 133:47–59

28. Griffith WP (1969) Raman spectroscopy of minerals. Nature 224(5216):264

29. Guedes A, Valentim B, Prieto AC, Noronha F (2012) Raman spectroscopy of coal macerals and fluidized bed char morphotypes. Fuel 97:443–449

30. Guedes A, Valentim B, Prieto AC, Rodrigues S, Noronha F (2010) Micro-Raman spectroscopy of collotelinite, fusinite and macrinite. Int J Coal Geol 83(4):415–422

31. Guedes A, Noronha F, Carmelo Prieto A (2005) Characterisation of dispersed organic matter from lower Palaeozoic metasedimentary rocks by organic petrography, X-ray diffraction and micro-Raman spectroscopy analyses. Int J Coal Geol 62(4):237–249

32. Hackley PC, Cardott BJ (2016) Application of organic petrography in North American shale petroleum systems: a review. Int J Coal Geol 163:8–51

33. Hackley PC, Lewan M (2018) Understanding and distinguishing reflectance measurements of solid bitumen and vitrinite using hydrous pyrolysis: implications to petroleum assessment. AAPG Bull 102(6):1119–1140

34. Hackley PC, Araujo CV, Borrego AG, Bouzinos A, Cardott BJ, Cook AC, Eble C et al (2015) Standardization of reflectance measurements in dispersed organic matter: results of an exercise to improve interlaboratory agreement. Marine Petrol Geol 59:22–34

35. Hampartsoumian E, Nimmo W, Rosenberg P, Thomsen E, Williams A (1998) Evaluation of the chemical properties of coals and their maceral group constituents in relation to combustion reactivity using multi-variate analyses. Fuel 77(7):735–748

36. Haskin LA, Wang A, Rockow KM, Jolliff BL, Korotev RL, Viskupic KM (1997) Raman spectroscopy for mineral identification and quantification for in situ planetary surface analysis: a point count method. J Geophys Res Planets 102(E8):19293–19306

37. Henry DG, Jarvis I, Gillmore G, Stephenson M (2019) A rapid method for determining organic matter maturity using laser Raman spectroscopy: application to carboniferous organic-rich mudstones and coals. Int J Coal Geol

38. Hinrichs R, Brown MT, Vasconcellos MAZ, Abrashev MV, Kalkreuth W (2014) Simple procedure for an estimation of the coal rank using micro-Raman spectroscopy. Int J Coal Geol 136:52–58

39. Hooghan K, Hathon L, Dixon M, Myers M (2017) Organic matter characterization in shales: a systematic empirical protocol. Microsc Microanal 23(S1):2130–2131

40. Houseknecht DW, Bensley DF, Hathon LA, Kastens PH (1993) Rotational reflectance properties of Arkoma Basin dispersed vitrinite: insights for understanding reflectance populations in high thermal maturity regions. Org Geochem 20(2):187–196

41. Hower JC, O'Keefe JMK, Watt MA, Pratt TJ, Eble CF, Stucker JD, Richardson AR, Kostova IJ (2009) Notes on the origin of inertinite macerals in coals: observations on the importance of fungi in the origin of macrinite. Int J Coal Geol 80(2):135–143

42. Hutton A, Bharati S, Robl T (1994) Chemical and petrographic classification of kerogen/macerals. Energy Fuels 8(6):1478–1488

43. ISO-7404/5.2009 (2009) Methods for the petrographic analysis of coals. Part 5: method of determining microscopically the reflectance of vitrinite

44. Jacob H (1989) Classification, structure, genesis and practical importance of natural solid oil bitumen ("migrabitumen"). Int J Coal Geol 11(1):65–79
45. Kelemen SR, Fang HL (2001) Maturity trends in Raman spectra from kerogen and coal. Energy Fuels 15(3):653–658
46. Kelemen SR, Afeworki M, Gorbaty ML, Sansone M, Kwiatek PJ, Walters CC, Freund H et al (2007) Direct characterization of kerogen by X-ray and solid-state 13C nuclear magnetic resonance methods. Energy Fuels 21(3):1548–1561
47. Kelemen SR, Walters CC, Kwiatek PJ, Freund H, Afeworki M, Sansone M, Lamberti WA et al (2010) Characterization of solid bitumens originating from thermal chemical alteration and thermochemical sulfate reduction. Geochim Cosmochim Acta 74(18):5305–5332
48. Khatibi S, Abarghani A, Liu K, Guedes A, Valentim B, Ostadhassan M (2020) Backtracking to parent maceral from produced bitumen with Raman spectroscopy. Minerals 10(8):679
49. Khatibi S, Ostadhassan M, Aghajanpour A (2018) Raman spectroscopy: an analytical tool for evaluating organic matter. J Oil Gas Petrochem Sci 1(1):28–33
50. Khatibi S, Ostadhassan M, Tuschel D, Gentzis T, Carvajal-Ortiz H (2018) Evaluating molecular evolution of Kerogen by Raman spectroscopy: correlation with optical microscopy and Rock-Eval pyrolysis. Energies 11(6):1406
51. Khatibi S, Ostadhassan M, Tuschel D, Gentzis T, Bubach B, Carvajal-Ortiz H (2018) Raman spectroscopy to study thermal maturity and elastic modulus of kerogen. Int J Coal Geol 185:103–118
52. Khorasani GK, Michelsen JK (1993) The thermal evolution of solid bitumens, bitumen reflectance, and kinetic modeling of reflectance: application in petroleum and ore prospecting. Energy Sour 15(2):181–204
53. Kondla D, Sanei H, Embry A, Ardakani OH, Clarkson CR (2015) Depositional environment and hydrocarbon potential of the Middle Triassic strata of the Sverdrup Basin, Canada. Int J Coal Geol 147:71–84
54. Lafargue E, Marquis F, Pillot D (1998) Rock-Eval 6 applications in hydrocarbon exploration, production, and soil contamination studies. Revue de l'institut français du pétrole 53(4):421–437
55. Li Z, Fredericks PM, Ward CR, Rintoul L (2010) Chemical functionalities of high and low sulfur Australian coals: a case study using micro attenuated total reflectance–Fourier transform infrared (ATR–FTIR) spectrometry. Org Geochem 41(6):554–558
56. Lis GP, Mastalerz M, Schimmelmann A, Lewan MD, Artur Stankiewicz B (2005) FTIR absorption indices for thermal maturity in comparison with vitrinite reflectance R0 in type-II kerogens from Devonian black shales. Organic Geochem 36(11):1533–1552
57. Liu D, Xiao X, Tian H, Min Y, Zhou Q, Cheng P, Shen J (2013) Sample maturation calculated using Raman spectroscopic parameters for solid organics: methodology and geological applications. Chin Sci Bull 58(11):1285–1298
58. Liu Z, Hathon LA, Myers MT (2017) Characterization of thermal maturity and organic material type with Raman spectroscopy in shale reservoirs. In: SPWLA 58th annual logging symposium. Society of Petrophysicists and Well-Log Analysts
59. Mählmann RF, Frey M (2012) Standardisation, calibration and correlation of the Kübler-index and the vitrinite/bituminite reflectance: an inter-laboratory and field related study. Swiss J Geosci 105(2):153–170
60. Mählmann RF, Le Bayon R (2016) Vitrinite and vitrinite like solid bitumen reflectance in thermal maturity studies: correlations from diagenesis to incipient metamorphism in different geodynamic settings. Int J Coal Geol 157:52–73
61. Marqués M, Suárez-Ruiz I, Flores D, Guedes A, Rodrigues S (2009) Correlation between optical, chemical and micro-structural parameters of high-rank coals and graphite. Int J Coal Geol 77(3–4):377–382
62. Mastalerz M, Drobniak A, Stankiewicz AB (2018) Origin, properties, and implications of solid bitumen in source-rock reservoirs: a review. Int J Coal Geol
63. Mastalerz M, Marc Bustin R (1995) Application of reflectance micro-Fourier transform infrared spectrometry in studying coal macerals: comparison with other Fourier transform infrared techniques. Fuel 74(4):536–542

64. Mastalerz M, Marc Bustin R (1993) Electron microprobe and micro-FTIR analyses applied to maceral chemistry. Int J Coal Geol 24(1–4):333–345

65. Mendonça Filho JG (1999) Aplicação de estudos de palinofácies e fácies orgânica em rochas do Paleozóico da Bacia do Paraná, Sul do Brasil. Universidade Federal do Rio Grande do Sul. Tese de Doutorado 2

66. Morga R (2011) Micro-Raman spectroscopy of carbonized semifusinite and fusinite. Int J Coal Geol 87(3–4):253–267

67. Mumm AS, İnan S (2016) Microscale organic maturity determination of graptolites using Raman spectroscopy. Int J Coal Geol 162:96–107

68. Orr WL (1986) Kerogen/asphaltene/sulfur relationships in sulfur-rich Monterey oils. Org Geochem 10(1–3):499–516

69. Ostadhassan M, Liu K, Li C, Khatibi S (2018) Fine scale characterization of shale reservoirs: methods and challenges. Springer

70. Peters KE, Walters CC, Moldowan JM (2005) The biomarker guide. Volume 2: Biomarkers and isotopes in the environment and human history, vol 75. Cambridge University Press, Cambridge, p 76

71. Peters K (1986) Guidelines for evaluating petroleum source rock using programmed pyrolysis. AAPG Bull 70:318–329

72. Pickel W, Kus J, Flores D, Kalaitzidis S, Christanis K, Cardott BJ, Misz-Kennan M et al (2017) Classification of liptinite–ICCP system 1994. Int J Coal Geol 169:40–61

73. Rezaiyan J, Cheremisinoff NP (2005) Gasification technologies: a primer for engineers and scientists. CRC press

74. Rippen D, Littke R, Bruns B, Mahlstedt N (2013) Organic geochemistry and petrography of lower cretaceous Wealden black shales of the lower Saxony basin: the transition from lacustrine oil shales to gas shales. Org Geochem 63:18–36

75. Sauerer B, Craddock PR, AlJohani MD, Alsamadony KL, Abdallah W (2017) Fast and accurate shale maturity determination by Raman spectroscopy measurement with minimal sample preparation. Int J Coal Geol 173:150–157

76. Schito A, Romano C, Corrado S, Grigo D, Poe B (2017) Diagenetic thermal evolution of organic matter by Raman spectroscopy. Org Geochem 106:57–67

77. Schoenherr J, Littke R, Urai JL, Kukla PA, Rawahi Z (2007) Polyphase thermal evolution in the Infra-Cambrian Ara Group (South Oman Salt Basin) as deduced by maturity of solid reservoir bitumen. Org Geochem 38(8):1293–1318

78. Scott AC, Glasspool IJ (2007) Observations and experiments on the origin and formation of inertinite group macerals. Int J Coal Geol 70(1–3):53–66

79. Spackman W, Moses RG (1960) Nature and occurrence of ash forming minerals in anthracite. No. SR-22. Pennsylvania State Univ., University Park (USA). Department of Geology

80. Spötl C, Houseknecht DW, Jaques RC (1998) Kerogen maturation and incipient graphitization of hydrocarbon source rocks in the Arkoma Basin, Oklahoma and Arkansas: a combined petrographic and Raman spectrometric study. Org Geochem 28(9–10):535–542

81. Stach E (1982) Stach's textbook of coal petrology

82. Styan WBT, Bustin RM (1983) Petrography of some fraser river delta peat deposits: coal maceral and microlitho type precursors in temperate-climate peats. Int J Coal Geol 2(4):321–370

83. Suárez-Ruiz I, Flores D, Filho JGM, Hackley PC (2012) Review and update of the applications of organic petrology: Part 1, geological applications. Int J Coal Geol 99:54–112

84. Taylor GH, Teichmüller M, Davis ACFK, Diessel CFK, Littke R, Robert P (1998) Organic petrology

85. Teichmüller M (1989) The genesis of coal from the viewpoint of coal petrology. Int J Coal Geol 12(1–4):1–87

86. Tissot BP, Welte DH (1984) From kerogen to petroleum. In: Petroleum formation and occurrence. Springer, Berlin, pp 160–198

87. Tissot B, Welte D (2012) Petroleum formation and occurrence: a new approach to oil and gas exploration. Springer Science & Business Media

88. Trudinger PA, Walter MR, Ralph BJ (eds) (2013) Biogeochemistry of ancient and modern environments: proceedings of the fourth international symposium on environmental biogeochemistry (ISEB) and, conference on biogeochemistry in relation to the mining industry and environmental pollution (leaching conference), held in Canberra, Australia, 26 August–4 September 1979. Springer Science & Business Media
89. Tuschel D (2013) Raman spectroscopy of oil shale. Spectroscopy 28(3):20–27
90. Tyson RV (1995) Palynological kerogen classification. In: Sedimentary organic matter, pp 341–365. Springer, Dordrecht
91. Van Krevelen DW (1950) Graphical-statistical method for the study of structure and reaction processes of coal. Fuel 29:269–284
92. Vandenbroucke M, Largeau C (2007) Kerogen origin, evolution and structure. Org Geochem 38(5):719–833
93. Vincent AJ (1995) Palynofacies analysis of Middle Jurassic sediments from the Inner Hebrides. PhD dissertation, Newcastle University
94. Waples DW (2013) Geochemistry in petroleum exploration. Springer Science & Business Media
95. Waples DW (1981) Organic geochemistry for exploration geologists. Burgess Pub. Co.
96. Ward CR, Gurba LW (1999) Chemical composition of macerals in bituminous coals of the Gunnedah Basin, Australia, using electron microprobe analysis techniques. Int J Coal Geol 39(4):279–300
97. Wei L, Wang Y, Mastalerz M (2016)Comparative optical properties of vitrinite and other macerals from upper Devonian–Lower Mississippian New Albany Shale: implications for thermal maturity. Int J Coal Geol 168:222–236
98. Wilkins RWT, Diessel CFK, Buckingham CP (2002) Comparison of two petrographic methods for determining the degree of anomalous vitrinite reflectance. Int J Coal Geol 52(1–4):45–62
99. Wilkins RWT, Boudou R, Sherwood N, Xiao X (2014) Thermal maturity evaluation from inertinites by Raman spectroscopy: the 'RaMM' technique. Int J Coal Geol 128:143–152

Chapter 6
Structural Evolution of Organic Matter in Deep Shales by Spectroscopy (^{1}H & ^{13}C-NMR, XPS, and FTIR) Analysis

Abstract Amorphous organic matter in geomaterials also known as kerogen undergoes significant alteration in chemical structure during thermal maturation which is characterized using a combination of solid-state ^{1}H & ^{13}C-NMR, X-ray photoelectron spectroscopy (XPS), and Fourier transform infrared (FTIR) techniques. For this study, four kerogen samples (type-II) from the Bakken Formation were selected based on the differences in their thermal maturity, as well as representing the pre-oil and oil window stages as measured through organic petrology and bulk geochemical screening of the samples using programmed pyrolysis. Later, organic matter was extracted from selected aliquots for chemical spectroscopy. Results documented a systematic structural change in these four samples where the ratio of CH_3/CH_2 increased when the maturity increases, along with the presence of shorter aliphatic chain length. Furthermore, the aromatic carbon structure becomes more abundant in higher maturities and oil window stages quantified by ^{13}C-NMR, XPS, and FTIR. Also, the rate of increase in aromaticity demonstrates a considerable rift between pre-oil window and oil window stages, as verified through bulk geochemical screening of the samples with Rock–Eval 6 pyrolysis HI index. Notably, it's found that kerogen maturation causes the relatively bulky oxygen-related carbon compounds to reduce at the early stages of maturation (pre-oil window) followed by concentration of such compounds at higher maturity stages. Next, the ratio of carbonyl/carboxyl functional groups to aromatic carbon shows an increase in oil window stage while reduction of sulfur in higher maturities was detected mainly in the SO_x forms. Finally, nitrogen content of the samples is reported in a variety of forms which varied regardless of the thermal maturation. Its concluded that, the structural and chemical changes that occurs in the organic matter involves defunctionalization of heteroatom functional groups, coupled with an increase in cross-linked carbon in the residual remaining kerogen.

Keywords Shale · Organic compounds · Spectroscopy analysis

6.1   Introduction

The abundance of organic carbon and heteroatoms in shale layers, which are deposited in source rocks in the form of organic matter, leads to the generation of hydrocarbon through thermal maturation. The solid-state organic compound (known as kerogen) of these atoms breaks down and undergoes a significant structural and compositional transformation during thermal maturation [1, 2]. In this regard, the study pertaining to these changes in kerogen structure has the potential to better characterize CO_2 sequestration and the enhanced oil recovery (EOR) methods [3]. However, there is a significant difficulty in understanding the mechanism of kerogen cracking coupled to hydrocarbon generation due to chemical heterogeneity and complexity of the kerogen molecular structure. Moreover, the limitation of analytical methods could not thoroughly provide quantitative molecular information of kerogen structure [4].

Kerogen characterization methods such as visual kerogen analysis (VKA) and pyrolysis (bulk geochemical techniques), are limited to the determination of the type of kerogen, maturity level, and hydrocarbon generating potential. Hence, in order to obtain the detailed structural information of kerogen, a number of spectroscopic techniques have been deployed [5–7]. The methods mentioned are destructive in addition to the limited information they can provide, thus other non-destructive methods such as Fourier transform infrared (FTIR) spectroscopy, Raman spectroscopy, X-ray photoelectron spectroscopy (XPS), X-ray diffraction (XRD), and ^{1}H & ^{13}C solid-state nuclear magnetic resonance (^{13}C-NMR) have been utilized to obtain quantitative structural information from the kerogen [8–10]. These direct methods can examine the solid kerogen sample without imposing any alterations so that the results would enable studying the structure and assist in building macromolecular models of kerogen though computational technique at the same time [11, 12].

Organic carbon in source rocks has been investigated by the solid-state ^{13}C-NMR technique coupled with FTIR [13, 14]. These spectroscopic tools can provide the qualitative and quantitative information related to both the structural assignment of carbon and the distribution of aromatic/aliphatic carbons in organic compounds in the last decade [15–17]. For studying the organic heteroatoms (oxygen, nitrogen, sulfur), FTIR and Raman are common methods employed for the analysis of functional groups in complex organic solids. However, due to the limitations of the functional groups acquired from these techniques, quantitative analysis are not adequate to fully provide detailed structural information. In order to complement missing data, XPS has been utilized for evaluating the organic compounds by overcoming this inherent deficiency to quantify organic heteroatoms with acceptable results [18, 19].

In this study, four isolated (extracted) kerogens from the mineral matrix at four different stages of natural thermal maturity from the Bakken Shale formation, from pre-oil to oil window stages, were examined by a combination of non-destructive analytical methods (^{1}H & ^{13}C-NMR, XPS, and FTIR) for chemical and structural analysis. The aim of this study was to identify structural characteristics and evolution of solid-state organic compounds as it undergoes thermal maturity in nature and to discover these changes in terms of quantitative and qualitative information from a chemical structure point of view. Additionally, we investigated other aspects of thermal maturation that can help us explain processes that would generate hydrocarbons from the organic compounds in the source rock.

6.2 Materials and Methods

Solid-state organic compound, type II kerogen samples were extracted from four different wells in the Bakken formation which is one of the largest unconventional shale oil plays in North America and is currently being studied for unconventional CO_2 EOR and sequestration [20–23]. Initially, the degree of maturity four Bakken kerogen samples (sample A-D) were examined by bitumen reflectance (%SBR_o) along with bulk geochemical properties using Rock-Eval 6 pyrolysis that is summarized in Table 6.1. In Fig. 6.1 and Table 6.1, samples A-B are at the thermally immature (pre-oil window) stage, while, samples C-D are at the peak mature (oil window) stage. It's important to note that selected kerogen samples were isolated using HCl and HF, and details of sample preparation and organic matter extraction procedure can be found in Abarghani et al. [23], and Khatibi et al. [24].

6.2.1 Solid-state ^1H & ^{13}C-NMR (Nuclear Magnetic Resonance)

Solid-state ^1H & ^{13}C-NMR analysis was carried out using a Bruker Avance III HD spectrometer with cross polarization-magic angle spinning (CP/MAS). The kerogen sample (100–150 mg) was packed and spun at 5 kHz. A contact time of 1.5 ms and recycle delay time of 5 s were used in the cross-polarization (CP). Resonance frequencies of ^1H-NMR and ^{13}C-NMR were 500 MHz and 125 MHz, respectively with the spectral widths of 10 kHz and 25 kHz for ^1H-NMR and ^{13}C-NMR. The relative proportion of different carbon types in the samples was quantified through curve fitting of the ^{13}C-NMR spectrum, which was conducted with the ratios of

Table 6.1 Properties of four solid-state organic compound (kerogen) samples (type II)[a]

Sample No.	T_{max} (°C)	TOC (wt%)	HI (mg/g C)	SBRo (%)
A	428	14.56	569	0.33
B	432	15.76	531	0.49
C	449	12.69	260	0.72
D	452	16.36	171	0.94

[a]The values were examined by the UV reflectance (%SBR_o) and Rock-Eval 6 with parameters: Tmax, TOC (Total organic carbon), HI (hydrogen index)

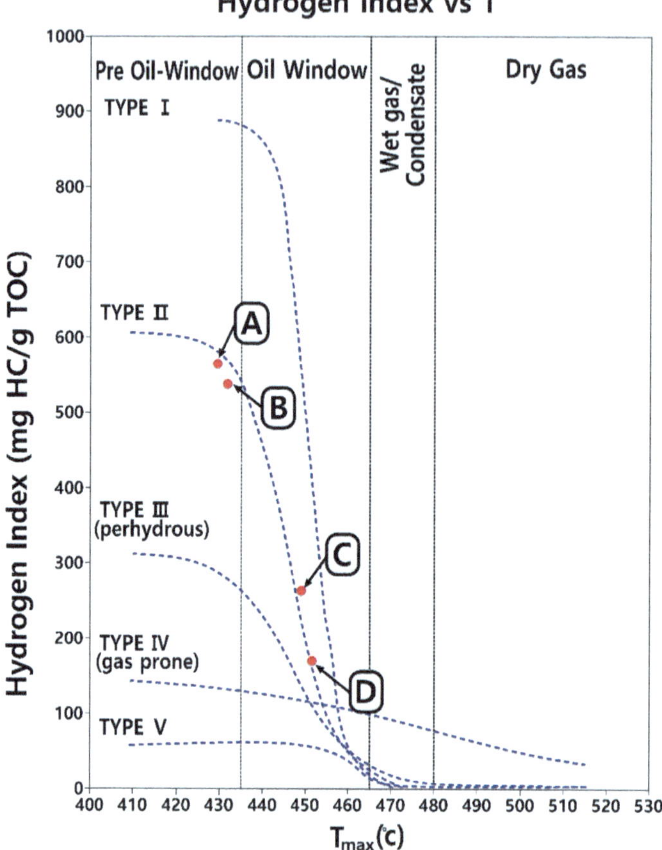

Fig. 6.1 Selected samples A-D are following kerogen type II thermal maturity trend where samples A-B are at the thermally immature (pre-oil window) stage, and samples C-D are at the peak mature (oil window) stage

Gaussian to Lorentzian distribution; the full width at half-maximum (fwhm). The details of structural parameters and assignments of chemical data shifts are presented in Table 6.2 [25, 26].

6.2.2 XPS (X-Ray Photoelectron Spectroscopy)

Freshly-powdered samples were pressed on indium foils and analyzed by a Al Kα (1486.6 eV) X-ray source with a pass energy of 89.5 eV and 44.75 eV for the survey and higher-resolution scans, respectively. During the XPS measurements, the pressure was kept at or below 1×10^{-9} mbar. The angle between the X-ray source,

Table 6.2 Structural parameters and assignments from ^{13}C-NMR spectra of the samples[a]

	f_{al}^{1}	f_{al}^{2}	f_{al}^{H1}	f_{al}^{H2}	f_{al}^{O1}	f_{a}^{O2}	f_{ar}^{H}	f_{ar}^{B}	f_{ar}^{S}	f_{ar}^{P}	f_{a}^{C}
ppm	(0–16)	(16–25)	(25–36)	(36–51)	(51–75)	(75–90)	(90–129)	(129–137)	(137–150)	(150–165)	(165–220)
A	0.080	0.050	0.240	0.20	0.056	0.02	0.23	0.032	0.047	0.015	0.030
B	0.075	0.064	0.196	0.19	0.057	0.03	0.25	0.035	0.052	0.024	0.027
C	0.070	0.067	0.165	0.10	0.049	0.05	0.34	0.037	0.061	0.029	0.032
D	0.056	0.066	0.120	0.10	0.051	0.04	0.37	0.045	0.084	0.030	0.038

[a] All ^{13}C chemical shift (ppm) were measured and assigned in carbon structural functionalities (f_{al}^{1}: aliphatic methyl, f_{al}^{2}: aromatic methyl, f_{al}^{H1}: methylene, f_{al}^{H2}: quaternary carbon, f_{al}^{O1}: methoxyl/aromatic methoxyl, f_{al}^{O2}: aliphatic carbon bonded to oxygen in cyclic hydrocarbon, f_{ar}^{H}: protonated aromatic carbon, f_{ar}^{B}: bridgehead aromatic carbon, f_{ar}^{S}: branched aromatic carbon, f_{ar}^{P}: oxy-aromatic carbon, and f_{a}^{C}: carbonyl/carboxyl carbon) [25, 26]

which is aligned along the surface normal, and the spectrometer is 54.7. All the XPS core-level spectra were analyzed using Augerscan and Origin software's. The core-level peaks are fitted using a Gaussian-Lorentzian (GL) function to include the instrumental response function along with the core-level line shape. The secondary electron background was subtracted using a Shirley function. We compensated the charging of the sample by irradiating the sample with an electron-flood gun (5 eV) [27].

6.2.3 *FTIR (Fourier Transform Infrared Spectroscopy)*

The infrared spectra of kerogen samples were recorded in adsorption range between 450 and 4000 cm^{-1} using a Thermo Fisher Scientific, Nicolet iS50 FTIR Spectrometer. Four kerogen samples were pulverized using a ball mill prior to characterizations. Fourier transform infrared (FTIR) spectroscopy using attenuated total reflectance (ATR) was employed to analyze kerogen extracts. Unlike transmittance FTIR, ATR does not require that kerogen samples be mixed with potassium bromide and formed into pellets under high pressure, thus reducing the time needed for the sample preparation. Kerogen samples were placed in contact with an internal reflection compound and IR spectra were obtained based on the excitation of the molecular vibrations of chemical bonds by the absorption of the light. The stretching absorptions of a vibrating chemical bond are observed at higher frequencies (wavenumbers) than the corresponding bending or bond deformation vibrations [28]. Because kerogen is rather a complex entity which limits the analysis of integrated bands in the bending vibration area, we decided to look at the C–H set of stretch vibrations observed in kerogen samples, as proposed in previous studies [28, 29], bands of the absorption were identifying by comparison with published spectra.

6.3 Results

6.3.1 *Solid-State ^{1}H & ^{13}C-NMR*

Aromatic and aliphatic proton regions of ^{1}H-NMR spectrum provide relatively qualitative information of hydrogen at different maturity levels [26]. From the acquired spectra in Fig. 6.2 (left), aromatic and aliphatic protons exist in the 6.4–8.3 ppm range and 0.5–4.3 ppm range, respectively. The ^{1}H-NMR spectra of sample A–D in Fig. 6.2 (left) show a broad asymmetric line centered around 0–1 ppm (centered at 0.64, 0.24, 0.54, and 0.91 ppm respectively). Sample A (immature) has a strong intensity in the range of the aliphatic protons between 0.5 to 4.3 ppm; the peak at 4.15 ppm indicates that the ratio of aliphatic protons to the aromatic ring is higher

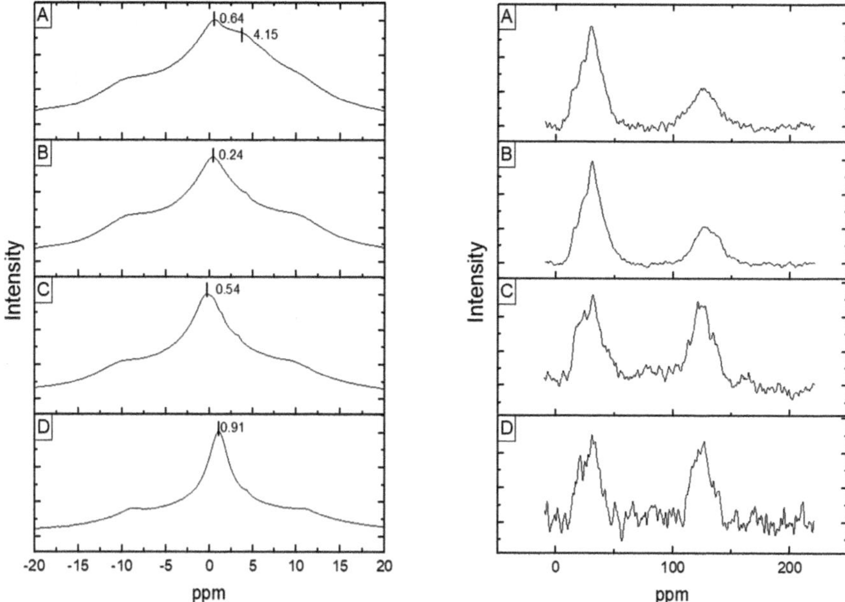

Fig. 6.2 ^{1}H-NMR (left) and ^{13}C-NMR (right) spectra of kerogen samples of samples from A–D

than the rest of the samples. In this regard, the results reveal that the shape of ^{1}H-NMR spectrum becomes sharp as the degree of maturity increases, which manifests the number of protons attached to aliphatic carbon reduces. However, the ^{1}H-NMR spectra output as a result of highly overlapping peaks was difficult to distinguish and quantify the structural formation of hydrogen which has been a problem in the past [30, 31]. Therefore, we decided to rely solely on the result of ^{13}C-NMR to extract quantitative information about carbon related structures.

The ^{13}C-NMR spectra of kerogen samples mainly reveal three regions: the aliphatic region at 0–90 ppm, the aromatic region at 100–165 ppm, and carbonyl/ carboxyl carbon region at 165–220 ppm [25, 26]. Figure 6.2 (right) illustrates that the ^{13}C-NMR spectrum of immature kerogen is distinct from the mature one in the range of the aromatic carbons between 90 to 220 ppm. Highly matured kerogen has a stronger intensity in ^{13}C–NMR spectrum related to aromatic structure. To acquire the details of carbon structural information, overlapping peak resolving of the ^{13}C-NMR spectra are deconvoluted by the fitting procedure. The relative areas calculated by the peak fitting, which represent the relative amount of the carbon-containing functional groups, are listed in Table 6.2.

As shown in the ^{13}C-NMR analysis (Table 6.2), with increasing thermal maturity, the aliphatic functional groups (0–90 ppm) exhibit a reduction trend in the relative amount of carbon, whereas the aromatic carbon groups (90–165 ppm) increase. This behavior is in an agreement with ^{1}H-NMR spectra in Fig. 6.2 (left). The nonpolar alkyl carbons (0–51 ppm) generally diminish with the degree of maturity, except the aromatic methyl ($f_{al}{}^2$) group which could be reflected by the increase of aromatic carbon. In addition, we infer that loss of oxygen at the early stages of maturation is due to a comparatively weak bond between oxygen–related carbon and kerogen backbone [19]. However, the results revealed that there isn't any meaningful relationship between oxygen–related carbons ($f_{al}{}^{O1}, f_{al}{}^{O2}, f_a{}^C$) and thermal maturity. We understand that, because the stoichiometry of oxygen–related signal is uncertain [29], it is difficult to estimate the accurate values concerning oxygen–related carbon. However, this study additionally investigated the changes of oxygen–related carbon with XPS and FTIR during the maturation, and based on the results given in Table 6.2, the evaluations of the carbon skeletal structure were correlated to the maturity levels of kerogen that is later discussed.

6.3.2 XPS

Different types of organic carbon and oxygen forms were determined by analyzing the carbon (1s) peak. In each sample, we determined four peaks located approximately at 285, 286, 287.5, and 289 eV after fitting the XPS carbon (1s) signal. The peak at 285 was attributed to both aromatic and aliphatic carbon. The amount of aromatic carbon was estimated using the XPS technique for $\prod$ to $\prod^*$ signal intensity as has been used previously [32]. This confirms that the highly mature kerogen has an abundance of aromatic carbon structure as revealed in the solid-state ^{1}H & ^{13}C-NMR results. The peaks at 286, 287.5, and 289 eV originate from carbon atoms bound to one oxygen atom by a single bond (C–O), carbon atoms are bound to one oxygen atom by double bonds (C=O), and carbon atoms are bound to two oxygen atoms via both double and single bonds (O=C–O), respectively. In the most mature stage (sample D), the results show a relatively large amount of enrichment for oxygen-related carbon, whereas the rest of the samples (A–C) exhibit a comparatively similar amount of oxygen-related carbons in kerogen structure (Table 6.3).

Organic nitrogen and sulfur information in kerogen structure was also obtained using XPS curve fitting methods, which at different energy positions are used to fit with the XPS N 1s and S 2p spectra, respectively. Nitrogen (1s) kerogen spectra were curve-resolved using four peaks at fixed energy positions of 397, 398.6, 399.4, 400.2, and 401.4 eV. Considering the organic sulfur (2p), the binding energy between 162–165.7 eV is assigned to pyrite, aliphatic and aromatic sulfur, and sulfoxide in XPS S 2p spectrum having 2p$_{3/4}$ and S 2p$_{1/2}$. The peaks at 168.0 ($\pm$ 0.5) to 170.5 ($\pm$ 0.5) eV can be indexed to S 2p$_{3/4}$ and S 2p$_{1/2}$ of SO$_x$ (sulfate/sulfite/sulfone). The peak locations, which are a little higher than the expectation, could be affected by the poor electrical contact in surface oxidation; the accumulation of positive charge shifts peaks toward higher binding energy. Because the peaks at 168.0 and 160.5 eV are overlapped and have also been observed in previous literature, we consider that the summation of this region presents the SO$_x$ assignment of sulfate, sulfite, and

Table 6.3 Structural parameters and assignment for the XPS spectra of the kerogen samples

Elemental	Functionality	Binding energy (eV)	Mole percent			
			A	B	C	D
C1 1s	Aliphatic	248.8	49	58	40	16.4
	Aromatic[a]		39	34	50	55
	C–O	286	1.2	0.6	1.4	23.1
	O–C=O	287.5	3.5	6.2	11	6.2
	C=O	289	7.3	4.6	2.8	2.3
N 1s	Nitride	397	–	19.8	24.8	12.5
	Pyridinic	398.6	–	16.4	15.7	22.2
	Quaternary	401.4	–	23.6	10.6	31.3
	Amino	399.4	–	20.6	42	9.8
	Pyrrolic	400.2	–	19.6	6.8	24.2
S 2p	Pyrite (FeS_2)	162	13.5	16.4	6.2	0.5
	Aliphatic	163	10.1	1.7	1.5	11.9
	Aromatic	164	20.1	1.3	3	14.9
	Sulfoxide	165.7	0.3	0	0	5.3
	SO_x[b]	168–170.5	58.0	80.6	89.4	67.4

[a]The amount of aromatic carbon was estimated by the technique of XPS Π to Π^* signal intensity in previous literature [32]
[b]The summation of the region presents the SOx assignment of sulfate, sulfite, and sulfoxide

sulfoxide [29]. Table 6.3 exhibits the details of quantitative structural parameters based on the curve fitting into different components of the selected kerogen samples (A–D).

6.3.3 FTIR

The intensities, depicted in Fig. 6.3, correspond to the asymmetric and symmetric stretching and bending of the C–H/C=O/C=C–C bonds about the central carbon atom on a relative basis in the spectra of kerogen samples. The reason for the overlapped original spectra is due to the same amount of energy, which is required for several vibrations. Therefore, IR structural evaluations have been established from acquired spectra documented in earlier studies [26]. To find the intensities at desired frequencies, the FTIR spectrum area 1800–1500 cm^{-1} and 3200–2700 cm^{-1} was fitted by the Fourier self-deconvolution [33], the coefficients of determination (R^2) of the peak fitting for the region in all spectra were found between 0.994–0.996.

Here, we examined the C–H stretching band intensities for CH_3, CH_2, aromatic ring CH, aromatic ring C=C–C, and C=O based on the deconvolution results. Four indices (CH_3/CH_2 ratio, aromaticity, A-factor, and C-factor) were used to evaluate the

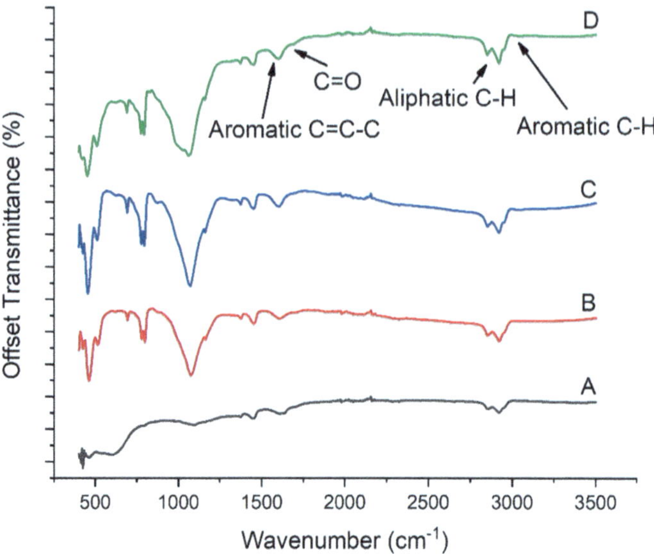

Fig. 6.3 FTIR spectra of kerogens with increasing maturity exhibit the changes in the C–H stretch vibration area (2800–3200 cm^{-1}), aromatic C=C–C ring stretch vibration area (1615–1580 cm^{-1}), and carbonyl/carboxyl group vibration area (1650–1750 cm^{-1})

structural characteristic of each different kerogen. The CH_3/CH_2 ratio [$I_{(2970-2950)}$/$I_{(2935-2915)}$] indicates the average chain length of aliphatic and the degree of chain branching. Aromaticity [$I_{(3130-3070)}/(I_{(2970-2950)} + I_{(2935-2915)})$] represents the degree of aromatic structure versus aliphatic chain structure. The A-factor [$I_{(2935-2915)}$/$(I_{(2935-2915)} + I_{(1615-1580)})$] and C-factor [$I_{(1750-1650)}/(I_{(1750-1650)} + I_{(1615-1580)})$] describe the relative amount of aliphatic carbon and oxygenated carbon to aromatic carbon, respectively [26] and all indices are calculated through integrating the area under the curves. When the maturity increases, the CH_3/CH_2 ratio acquired demonstrates that the aliphatic chain length is comparatively shorter, and the branching is developed. Restated, highly mature kerogen has more aromatic ring contribution and less methyl/methylene involvement with the expected increase in aromatic structure (aromaticity index). The observed trends corroborate findings from previous studies where, during the maturation, chemical structure changes have been drawn into shorten aliphatic chain length and expanding aromaticity [26]. The result of A-factor, regarding the relative amount of aliphatic to aromatic carbon, indicates that there is generally a decreasing trend over the degree of maturation. Oxygen-related information (C-factor) reveals that the ratio of oxygen-related carbon to aromatic carbon initially decreases (immature stage), then rises at the mature stage advances; the comparative details of the indices are discussed in more details in the discussion section.

6.4 Discussion

6.4.1 Carbon Structural Changes

The ratios of CH_3/CH_2 from ^{13}C-NMR have a similar order of increase to those obtained from the FTIR analysis when the maturity increases (Fig. 6.4a–b), which is consistent with the previously described data from type II kerogen structure [34]. This growth of CH_3/CH_2 ratio could be evidence of chain cleavage coupled with hydrocarbon generation. While kerogen maturation occurs naturally, the molecular structure of this macromolecule evolves through bond-breaking and bond-forming, under changing physical conditions (temperature, pressure, and time) of the subsurface [35]. In terms of carbon to carbon (C–C) bond in the kerogen structure, bond-dissociation energy (BDE) between α and β carbon is weaker than one between α and aromatic carbon [33, 36]. Also, short branching carbon (such as methyl) to carbon in the backbone structure has less BDE than longer chain carbon [34]. It appears that kerogen samples of the Bakken shale play also follow this mechanism which indicates a higher ratio of CH_3/CH_2 during thermal maturation as shown in Fig. 6.4a, b. In addition to the ratio of CH_3/CH_2, the average aliphatic carbon length and the amount of aromatic carbon follows this trend too. Figure 6.4c explains that the average length of aliphatic carbon reduces as the maturity increases meaning thereby that thermal maturation accompanies the structural changes in kerogen with both shorter aliphatic chain and a higher ratio of CH_3/CH_2, as delineated by FTIR and ^{13}C-NMR.

Also, the A-factor calculated through FTIR data (Fig. 6.4d) shows that the relative ratio of aliphatic to aromatic carbon generally decreases in accordance with the shape of the ^{1}H-NMR spectrum apart from sample B that has the highest overall value. The higher A-factor value of sample A than sample B was expected because the ratio of CH_3/CH_2 increases and as a result aliphatic chain length becomes shorter. However, sample B was found to have the largest A-factor, and this could be explained by the fact that the quantities of aromatic carbon in sample B is relatively lower than was expected. This result is confirmed by the information that is obtained from the analysis of aromatic carbon that is presented in Fig. 6.5.

Figure 6.5 depicts the aromatic carbon as investigated by three different analytical methods, ^{13}C-NMR, FTIR, and XPS. Figure 6.5a–c verifies our finding that higher maturity kerogen overall tends to have more amount of aromatic carbon. This is also compatible with the ^{1}H-NMR result, which infers to the ratio of aliphatic protons to the aromatic ring compared to the rest of protons. With an increase of maturity (sample A to D), the spectrum becomes sharper and has diminished intensity in the range of the aliphatic protons between 0.5–4.3 ppm, which represents the abundance of aromatic structure. This observation is collectively acceptable excluding sample B with the amount of aromatic carbon of XPS being comparatively 13% lower than the one of ^{13}C-NMR.

FTIR results also propose that the aromaticity value of sample B is similar to sample A, which is different than the overall observed trend. This latest outcome

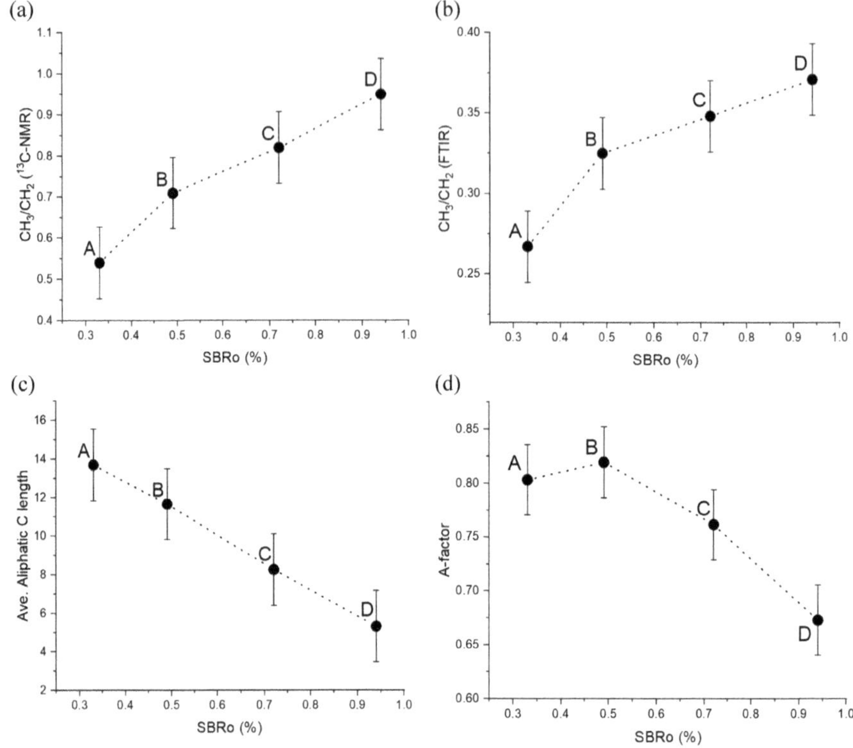

Fig. 6.4 The aliphatic carbon structure changes by thermal maturity. The ratio of methyl/methylene is estimated from **a** ^{13}C-NMR [$(f_{al}^{1} + f_{al}^{2})/f_{al}^{H1}$] and **b** FTIR [$I_{(2970-2950)}/I_{(2935-2915)}$] results. **c** Average aliphatic carbon chain length by ^{13}C-NMR [$(f_{al}^{1} + f_{al}^{2} + f_{al}^{H1} + f_{al}^{H2} + f_{al}^{O1} + f_{al}^{O2})/f_{ar}^{S}$]. **d** A-factor by FTIR [$I_{(2935-2915)}/(I_{(2935-2915)} + I_{(1615-1580)})$]. Error bars represent standard errors

can support lower A-factor values for Sample B compared to Sample A, as seen in Fig. 6.4d. In addition, the degree of aromaticity (Fig. 6.5d) estimated from three techniques (^{13}C-NMR, FTIR, and XPS) strongly confirms that the aromaticity is increasing. Notably, we found that based on the aromaticity results there is relatively a considerable rift between pre-oil and oil window stage samples. Following the oil window stage, the amount of aromatic carbon fairly increases, as well as the rate of change in aromaticity as seen by the slope of the curve which is found higher through advancing thermal maturity in the oil generation window.

Based on the results, the structural changes of the Bakken kerogen, coupled with hydrocarbon generation, are affected by the C-C BDE during the thermal maturation process. Due to the weaker BDE between α–β carbons than α-aromatic carbons, breaking the aliphatic chain bond leads to both the hydrocarbon generation and the aromatic abundant structure in kerogen backbone. In other words, during the thermal maturation, the comparatively light hydrocarbon was detached, and the remaining

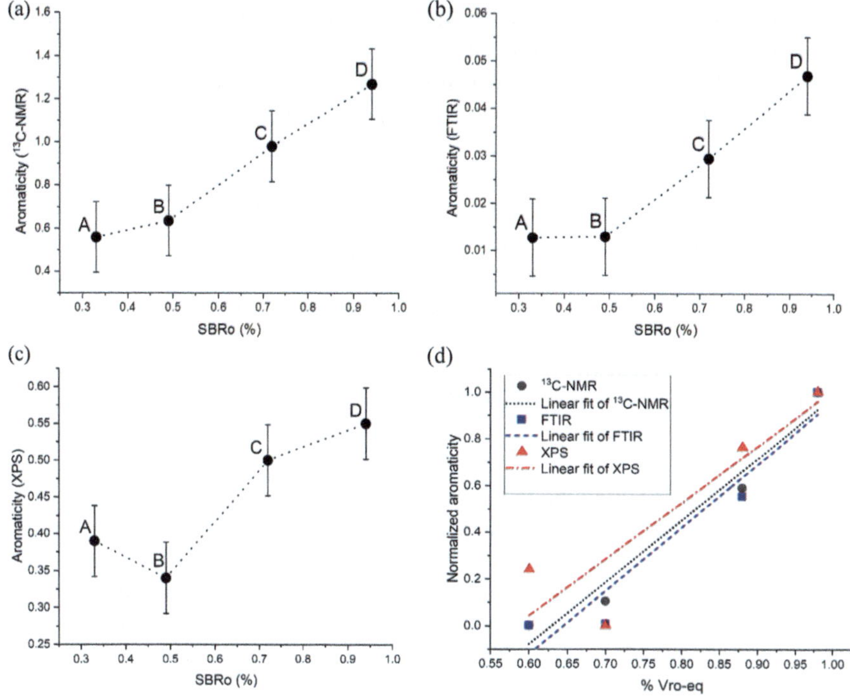

Fig. 6.5 The aromatic carbon structural changes by thermal maturity. The aromaticity (aromatic carbon/aliphatic carbon) was calculated from **a** ^{13}C-NMR $[f_{ar}^H + f_{ar}^B + f_{ar}^S + f_{ar}^P]$, **b** FTIR $[I_{(3130-3070)}/(I_{(2970-2950)} + I_{(2935-2915)})]$, and **c** XPS [mole % aromatic carbon]. **d** Comparison of aromaticity slop trends (linearly fitted). Error bars represent standard errors

kerogen backbone became a relatively solid-state organic compound having abundant aromatic structure and well organized rings.

6.4.2 Heteroatoms

The loss of oxygen functional groups drives early structural changes. Because the oxygen-related carbon bonds have relatively weak BDE, the carboxyl and carbonyl groups can be easily defunctionalized [34]. In our study, the results revealed that kerogen maturation involves relatively bulky oxygen-related carbon reduction at the early stage of maturation (pre-oil window) then it increases in oil window stage, which is verified by ^{13}C-NMR and XPS in Tables 6.2 and 6.3.

The changes in the ratios of C=O/aromatic carbon were investigated by three methods (^{13}C-NMR, FTIR, and XPS) depicted in Fig. 6.6. Because it is difficult to analyze the organic oxygen group from ^{13}C-NMR [29], the comparison of the three methods is presented in the same graph to help us better understand and verify the results. As thermal maturity increases, first, the ratio of C=O/aromatic carbon decreases regardless of the method, along with the ratio of carbonyl/carboxyl functional groups to aromatic carbon which also was found to reduce at the early stages of maturation. Then, in the oil-window stage (sample C–D), the ratio of C=O/aromatic carbon starts to rise showing that the amount of oxygen estimated in sample D to become higher than sample C. Although most oxygen-containing functional groups will be expelled from the organic matter during the initial stages of thermal maturation, the results exhibit an unexpected increase in the oxygen content of the functional groups in the higher maturity ranges of kerogen. This can be explained by the enrichment of oxygen through the addition of oxygen from inorganic mineral or water (interstitial water) impurities as thermal maturity increases which are also observed in other studies [34]. It's also expected that the formation of insoluble pyro-bitumen, which is highly oxidized as thermal maturity advances, may lead to an unexpected increase of C=O/aromatic carbon ratio in higher thermal maturation [8].

The observations pertaining to the organic sulfur in our kerogen samples clearly exhibits that the amount of sulfur would decrease significantly as maturation progresses even from the early onset of this process. The results that is obtained from the XPS data analysis depicts that sample A contains the largest number of sulfur atoms (2.8 sulfur atom per 100 carbon atom), whereas sample B–D were measured with lower quantities of sulfur (1.4, 1.1, and 0.7 per 100 carbons, respectively) compared to sample A. This observed phenomenon is due to the fact that the maturation process entails the depolymerization of kerogen structure by breaking the weaker bonds (C–O and C–S) that'd lead to generating heavier soluble bitumen compounds as explained in the literature [37]. Organic sulfur in aliphatic and aromatic compounds,

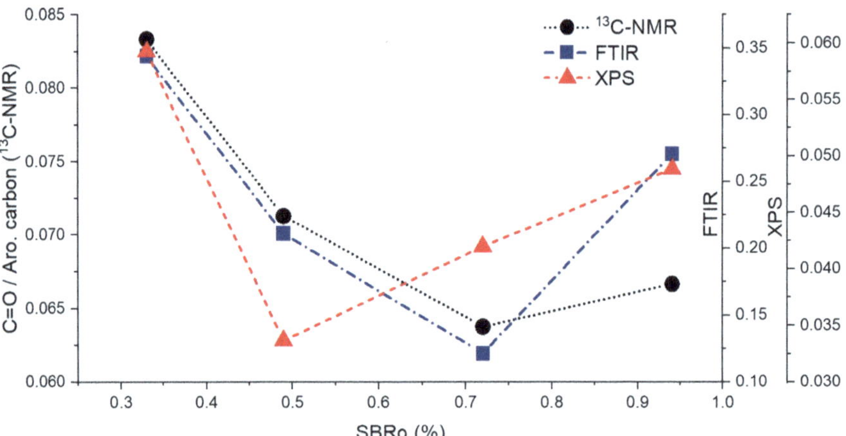

Fig. 6.6 The ratios of C=O/aromatic carbon were estimated from ^{13}C-NMR, FTIR, and XPS results, which were calculated by $f_a^C/(f_{ar}^H + f_{ar}^B + f_{ar}^S + f_{ar}^P)$, C-factor $[I_{(1650-1750)}/(I_{(1650-1750)} + I_{(1615-1580)})]$, and [(mole % of total oxygen) × (C=O and C–O=O fraction)/(mole % of aromatic carbon)], respectively

sulfoxide, and SO_x (sulfate/sulfite/sulfone) in addition to inorganic sulfur and pyrite is clearly recorded in the analysis and appear in all spectra as shown in Table 6.3. Organic nitrogen as detected by XPS revealed that it mainly contains pyrrolic, pyridinic, quaternary, and amino, which is detected by several peaks based on their energy positions [29]; the nitrogen data for each kerogen sample is summarized in Table 6.3. Based on the XPS results from the Bakken kerogen samples, we deduced that a variety of nitrogen containing compounds can exist in the samples regardless of their thermal maturation stage. This means, we were not able to establish a meaningful pattern in the changes that would occur in nitrogen quantities as a function of thermal maturation in our samples and further study is required to investigate the role of organic nitrogen during depolymerization of kerogen structure.

Generally speaking, based on the results, the weak bonds in the complex structure of kerogen start to detach containing bonds between C–O, and C–S at the early stages of maturation. These bond-breaking reactions would lead to the generation of H_2S gas, NSO compounds, and heavy soluble compound (bitumen). It needs to reiterate that the structural and chemical evolution of the Bakken kerogen involves defunctionalization of heteroatom functional groups, coupled with an increase in cross-linked carbon in the residual higher maturity kerogen.

6.5 Conclusions

This study shows how hydrocarbon generation can impose changes in the source rock organic compound structure during the maturation process. The analysis of extracted kerogen through spectroscopic techniques (^{1}H & ^{13}C-NMR, XPS, and FTIR) was used to determine chemical contents and molecular structures as maturation advances. This study will enable us to better model separation and capture processes in shale plays via theoretical methods. Based on the results the following conclusions can be made:

- The structural changes of kerogen pertaining to the abundance of the aromatic and aliphatic chain length correlate well with the maturity of kerogen. Based on the aliphatic carbon chain that is detected in kerogen structure by ^{1}H and ^{13}C-NMR and FTIR, the ratio of CH_3/CH_2 increases when the maturity increases accompanied with the shorter aliphatic chain length.
- In contrast to aliphatic carbon, the aromatic carbon structure becomes more abundant in higher maturity and oil window stage. All three techniques (^{13}C-NMR, XPS, and FTIR) confirms an increasing trend similar to aromaticity with the same rate of changes. Also, a considerable discrepancy between pre-oil window and oil window stage regarding aromaticity was observed.
- Kerogen maturation causes the relatively bulky oxygen compounds reduction at the early stage of maturation (pre-oil window) with an increase in the oil window stage, based on ^{13}C-NMR and XPS results. In addition, the ratio of carbonyl/carboxyl functional groups to aromatic carbon shows an increase in the

oil window stage (mature kerogen). This may be driven by contamination from inorganic oxygen in mineral or interstitial water impurities. Besides, the formation of insoluble oxygen-rich pyro-bitumen can lead to enrichment of the C=O/aromatic carbon ratio.

- The reduction of sulfur in kerogen was observed, as the level of maturity increases, mostly in SO_x forms. Finally, the XPS results exhibited that the diverse form of nitrogen could exist regardless of the thermal maturation stage, however, further investigations of heteroatoms are necessary for a better understanding of changes in NSO content as a function of thermal maturity.

References

1. Durand B (1980) Sedimentary organic matter and kerogen. Definition and quantitative importance of kerogen. Kerogen, insoluble organic matter from sedimentary rock, editions technips, pp 13–34
2. Rullkötter J, Michaelis W (1990) The structure of kerogen and related materials. A review of recent progress and future trends. Org Geochem 16(4–6):829–852
3. Lee H, Shakib FA, Shokouhimehr M, Bubach B, Kong L, Ostadhassan M (2019) Optimal separation of CO_2/CH_4/Brine with amorphous kerogen: a thermodynamics and kinetics study. J Phys Chem C 123(34):20877–20883
4. Vandenbroucke M, Largeau C (2007) Kerogen origin, evolution and structure. Org Geochem 38:719–833
5. Tong J, Han X, Wang S, Jiang X (2011) Evaluation of structural characteristics of Huadian oil shale kerogen using direct techniques (solid–state 13C NMR, XPS, FT–IR, and XRD). Energy Fuels 25:4006–4013
6. Tong L, Jiang X, Han X, Wang (2016) Evaluation of the macromolecular structure of Huadian oil shale kerogen using molecular modeling. Fuel 181:330–339
7. Clough A, Sigle JL, Jacobi D, Sheremata J, White JL (2015) Characterization of kerogen and source rock maturation using solid-state NMR spectroscopy. Energy Fuels 29:6370–6382
8. Lis GP, Mastalerz M, Schimmelmann A, Lewan MD, Stankiewicz BA (2005) FTIR absorption indices for thermal maturity in comparison with vitrinite reflectance R0 in type–II kerogens from Devonian black shales. Org Geochem 36:1533–1552. https://doi.org/10.1016/j.orggeo chem.2005.07.001
9. Khatibi S, Ostadhassan M, Tuschel D, Gentzis T, Bubach B, Carvajal-Ortiz H (2018) Raman spectroscopy to study thermal maturity and elastic modulus of kerogen. Int J Coal Geol 185:103–118
10. Wei Z, Gao Z, Zhang D, Da J (2005) Assessment of thermal evolution of kerogen geopolymers with their structural parameters measured by solid-state 13C NMR spectroscopy. Energy Fuels 19:240–250
11. Lee H, Ostadhassan M, Liu K, Bubach B (2019) Developing an amorphous kerogen molecular model based on gas adsorption isotherms. [Online early access]. https://doi.org/10.26434/chemrxiv.7965152. Published Online: 09 April 2019. https://chemrxiv.org/s/dbfe102258d971e948f6
12. Khatibi S, Ostadhassan M, Aghajanpour A (2018) Raman spectroscopy: an analytical tool for evaluating organic matter. J Oil Gas Petrochem Sci 1:28–33
13. Renault M, Cukkemane A, Baldus M (2010) Solid-state NMR spectroscopy on complex biomolecules. Angew Chem Int 49:8346–8357

14. Minor EC, Swenson MM, Mattson BM, Oyler AR (2014) Structural characterization of dissolved organic matter: a review of current techniques for isolation and analysis. Environ Sci Process Impacts 16:2064–2079
15. VanderHart DL, Retcofsky HL (1976) Estimation of coal aromaticities by proton–decoupled carbon-13 magnetic resonance spectra of whole coals. Fuel 55:202–204
16. Hatcher PG, Wilson MA, Vassallo AM, Lerch HE (1989) Studies of angiospermous wood in Australian brown coal by nuclear magnetic resonance and analytical pyrolysis: new insights into the early coalification process. Int J Coal Geol 13:99–126
17. Wang S, Tang Y, Schobert HH, Guo Y, Su Y (2011) FTIR and 13C NMR investigation of coal component of late Permian coals from Southern China. Energy Fuels 25:5672–5677
18. Kelemen SR, Afeworki M, Gorbaty ML (2002) Characterization of organically bound oxygen forms in lignites, peats, and pyrolyzed peats by X–ray photoelectron spectroscopy (XPS) and solid-state 13C NMR methods. Energy Fuels 16:1450–1462
19. Wang Q, Hou T, Wu W, Yu Z, Ren S, Liu Q, Liu Z (2017) A study on the structure of Yilan oil shale kerogen based on its alkali-oxygen oxidation yields of benzene carboxylic acids, 13C NMR and XPS. Fuel Process Technol 166:30–40
20. Lin R, Ritz GP (1993) Studying individual macerals using i.r. Microspectroscopy, and implications on oil versus gas/condensate proneness and "low–rank" generation. Org Geochem 20:695–706
21. Salmon E, Behar F, Lorant F, Hatcher PG, Marquaire PM (2009) Early maturation processes in coal. Part 1: pyrolysis mass balance and structural evolution of coalified wood from the morwell brown coal seam. Org. Geochem 40:500–509
22. Michels R, Langlois E, Ruau O, Mansuy L, Elie M, Landais P (1996) Evolution of asphaltenes during artificial maturation: a record of the chemical processes. Energy Fuels 10:39–48
23. Abarghani A, Ostadhassan M, Gentzis T, Carvajal-Ortiz H, Bubach B (2018) Organofacies study of the Bakken source rock in North Dakota, USA, based on organic petrology and geochemistry. Int J Coal Geol 188:79–93
24. Khatibi S, Ostadhassan M, Tuschel D, Gentzis T, Carvajai-Ortiz H (2018) Evaluating molecular evolution of kerogen by Raman spectroscopy: correlation with optical microscopy and Rock-Eval pyrolysis. Energies 11:1406
25. Wei Q, Tang T (2018) 13C–NMR study on structure evolution characteristics of high–organic-sulfur coals from typical Chinese areas. Minerals 8:49
26. Wang Q, Cui D, Wang P, Bai J, Wang Z, Liu B (2018) A comparison of the structures of > 300 °C fractions in six Chinese shale oils obtained from different locations using 1H NMR, 13C NMR and FT–IR. Fuel 211:341–352
27. Heide PVD (2012) X–ray photoelectron spectroscopy—an introduction to principles and practices. Wiley
28. Coates (2006) John interpretation of infrared spectra, a practical approach. Encycl Anal Chem. https://doi.org/10.1002/9780470027318.a5606
29. Kelemen SR, Afeworki M, Gorbaty ML, Sansone M, Kwiatek PJ, Walters CC, Freund H, Siskin M (2007) Direct characterization of kerogen by X–ray and solid-state 13C nuclear magnetic resonance methods. Energy Fuels 21:1548–1561
30. Borrego AG, Blanco CG, Prado JG, Diaz C, Guillen MD (1996) 1H NMR and FTIR spectroscopic studies od bitumen and shale oil from selected Spanish oil shales. Energy Fuels 10:77–84
31. Premovic PI, Jovanovic LS, Michel D (1992) Solid–state 13C and 1H NMR in kerogen research: uncertainty of aromaticity estimation. Appl Spectrosc 46:1750–1752
32. Kelemen SR, Rose KD, Kwiatek PJ (1993) Carbon aromaticity based on XPS π to π^* signal intensity. Appl Surf Sci 64:167–173
33. Yao X, Hou X, Wu G, Xu Y, Xiang H, Jiao H, Li Y (2002) Estimation of C-C bond dissociation enthalpies of large aromatic hydrocarbon compounds using DFT methods. J Phys Chem A 106:7184–7189
34. Craddock PR, Bake KD, Pomerantz AE (2018) Chemical, molecular, and microstructural evolution of kerogen during thermal maturation: case study from the Woodford Shale of Oklahoma. Energy Fuels 32:4859–4872

35. Takeda N, Asakawa T (1988) Study of petroleum generation by pyrolysis—I. Pyrolysis experiments by Rock–Eval and assumption of molecular structural change of kerogen using 13C–NMR. Appl Geochem 3:441–453
36. Manka MJ, Brown RL, Stein SE (1985) Flow ESR and static studies of the decomposition of phenyl-substituted ethanes. J Phys Chem 89:5421–5427
37. Behar F, Lorant F, Lewan M (2008) Role of NSO compound during primary cracking of a type II kerogen and a type III lignite. Org Geochem 39:1–22

Printed by Printforce, the Netherlands